The New Yor
Its Construction a

Anonymous

Alpha Editions

This edition published in 2022

ISBN: 9789356784802

Design and Setting By

Alpha Editions

www.alphaedis.com

Email - info@alphaedis.com

Contents

THE NEW YORK SUBWAY

OPERATING ROOM OF POWER HOUSE

INTERBOROUGH RAPID TRANSIT
The New York Subway

ITS CONSTRUCTION AND EQUIPMENT

NEW YORK

INTERBOROUGH RAPID TRANSIT COMPANY

ANNᴼ. DOMᴵ. MCMIV

INTERBOROUGH RAPID TRANSIT COMPANY

ALFRED SKITT
CORNELIUS VANDERBILT
GEORGE W. YOUNG

Executive Committee

AUGUST BELMONT
ANDREW FREEDMAN
JAMES JOURDAN
WALTER G. OAKMAN
WILLIAM A. READ
CORNELIUS VANDERBILT

Officers

AUGUST BELMONT, PRESIDENT
E. P. BRYAN, VICE-PRESIDENT
H. M. FISHER, SECRETARY
D. W. McWILLIAMS, TREASURER
E. F. J. GAYNOR, AUDITOR
FRANK HEDLEY, GENERAL SUPERINTENDENT
S. L. F. DEYO, CHIEF ENGINEER
GEORGE W. WICKERSHAM, GENERAL COUNSEL
CHAS. A. GARDINER, GENERAL ATTORNEY
DeLANCEY NICOLL, ASSOCIATE COUNSEL
ALFRED A. GARDNER, ASSOCIATE COUNSEL

Engineering Staff

S. L. F. DEYO, CHIEF ENGINEER.

Electrical Equipment

L. B. Stillwell, Electrical Director.
H. N. Latey, Principal Assistant.
Frederick R. Slater, Assistant Engineer in charge of Third Rail
Construction.
Albert F. Parks, Assistant Engineer in charge of Lighting.
George G. Raymond, Assistant Engineer in charge of Conduits and Cables.
William B. Flynn, Assistant Engineer in charge of Draughting Room.

Mechanical and Architectural

J. Van Vleck, Mechanical and Construction Engineer.
William C. Phelps, Assistant Construction Engineer.
William N. Stevens, Ass't Mechanical Engineer.
Paul C. Hunter, Architectural Assistant.
Geo. E. Thomas, Supervising Engineer in Field.

Cars and Signal System

George Gibbs, Consulting Engineer.
Watson T. Thompson, Master Mechanic.
J. N. Waldron, Signal Engineer.

RAPID TRANSIT SUBWAY CONSTRUCTION COMPANY

Directors

AUGUST BELMONT
E. P. BRYAN
ANDREW FREEDMAN
JAMES JOURDAN
GARDINER M. LANE
WALTHER LUTTGEN
JOHN B. MCDONALD
WALTER G. OAKMAN
JOHN PEIRCE
MORTON F. PLANT
WILLIAM A. READ
CORNELIUS VANDERBILT
GEORGE W. YOUNG

Executive Committee

AUGUST BELMONT
ANDREW FREEDMAN
JAMES JOURDAN
WALTER G. OAKMAN
WILLIAM A. READ
CORNELIUS VANDERBILT

Officers

AUGUST BELMONT, PRESIDENT
WALTER G. OAKMAN, VICE-PRESIDENT
JOHN B. MCDONALD, CONTRACTOR
H. M. FISHER, SECRETARY
JOHN F. BUCK, TREASURER
E. F. J. GAYNOR, AUDITOR
S. L. F. DEYO, CHIEF ENGINEER
GEORGE W. WICKERSHAM, GENERAL COUNSEL
ALFRED A. GARDNER, ATTORNEY

Engineering Staff

S. L. F. Deyo, Chief Engineer.

H. T. Douglas, Principal Assistant Engineer.

A. Edward Olmsted, Division Engineer, Manhattan-Bronx Lines.

Henry B. Reed, Division Engineer, Brooklyn Extension.

Theodore Paschke, Resident Engineer, First Division, City Hall to 33d Street, also Brooklyn Extension, City Hall to Bowling Green; and Robert S. Fowler, Assistant.

Ernest C. Moore, Resident Engineer, Second Division, 33d Street to 104th Street; and Stanley Raymond, Assistant.

William C. Merryman, Resident Engineer, Third Division, Underground Work, 104th Street to Fort George West Side and Westchester Avenue East Side; and William B. Leonard, W. A. Morton, and William E. Morris, Jr., Assistants.

Allan A. Robbins and Justin Burns, Resident Engineers, Fourth Division, Viaducts; and George I. Oakley, Assistant.

Frank D. Leffingwell, Resident Engineer, East River Tunnel Division, Brooklyn Extension; and C. D. Drew, Assistant.

Percy Litchfield, Resident Engineer, Fifth Division, Brooklyn Extension, Borough Hall to Prospect Park; and Edward R. Eichner, Assistant.

M. C. Hamilton, Engineer, Maintenance of Way; and Robert E. Brandeis, Assistant.

D. L. Turner, Assistant Engineer in charge of Stations.

A. Samuel Berquist, Assistant Engineer in charge of Steel Erection.

William J. Boucher, Assistant Engineer in charge of Draughting Rooms.

INTRODUCTION

The completion of the rapid transit railroad in the boroughs of Manhattan and The Bronx, which is popularly known as the "Subway," has demonstrated that underground railroads can be built beneath the congested streets of the city, and has made possible in the near future a comprehensive system of subsurface transportation extending throughout the wide territory of Greater New York.

In March, 1900, when the Mayor with appropriate ceremonies broke ground at the Borough Hall, in Manhattan, for the new road, there were many well-informed people, including prominent financiers and experienced engineers, who freely prophesied failure for the enterprise, although the contract had been taken by a most capable contractor, and one of the best known banking houses in America had committed itself to finance the undertaking.

In looking at the finished road as a completed work, one is apt to wonder why it ever seemed impossible and to forget the difficulties which confronted the builders at the start.

The railway was to be owned by the city, and built and operated under legislation unique in the history of municipal governments, complicated, and minute in provisions for the occupation of the city streets, payment of moneys by the city, and city supervision over construction and operation. Questions as to the interpretation of these provisions might have to be passed upon by the courts, with delays, how serious none could foretell, especially in New York where the crowded calendars retard speedy decisions. The experience of the elevated railroad corporations in building their lines had shown the uncertainty of depending upon legal precedents. It was not, at that time, supposed that the abutting property owners would have any legal ground for complaint against the elevated structures, but the courts found new laws for new conditions and spelled out new property rights of light, air, and access, which were made the basis for a volume of litigation unprecedented in the courts of any country.

An underground railroad was a new condition. None could say that the abutting property owners might not find rights substantial enough, at least, to entitle them to their day in court, a day which, in this State, might stretch into many months, or even several years. Owing to the magnitude of the work, delay might easily result in failure. An eminent judge of the New York Supreme Court had emphasized the uncertainties of the situation in the following language: "Just what are the rights of the owners of property abutting upon a street or avenue, the fee in and to the soil underneath the surface of which has been acquired by the city of New York, so far as the same is not required for the ordinary city uses of gas or water pipes, or others of a like character, has never been finally determined. We have now the example of the elevated railroad, constructed and operated in the city of New York under legislative and municipal authority for nearly twenty years, which has been compelled to pay many millions of dollars to abutting property owners for the easement in the public streets appropriated by the construction and maintenance of the road, and still the amount that the road will have to pay is not ascertained. What liabilities will be imposed upon the city under this contract; what injury the construction and operation of this road will cause to abutting property, and what easements and rights will have to be acquired before the road can be legally constructed and operated, it is impossible now to ascertain."

It is true, that the city undertook "to secure to the contractor the right to construct and operate, free from all rights, claims, or other interference, whether by injunction, suit for damages, or otherwise on the part of any abutting owner or other person." But another eminent judge of the same court had characterized this as "a condition absolutely impossible of fulfillment," and had said: "How is the city to prevent interference with the work by injunction? That question lies with the courts; and not with the courts of this State alone, for there are cases without doubt in which the courts of the United States would have jurisdiction to act, and when such jurisdiction exists they have not hitherto shown much reluctance in acting.... That legal proceedings will be undertaken which will, to some extent at least, interfere with the progress of this work seems to be inevitable...."

Another difficulty was that the Constitution of the State of New York limited the debt-incurring power of the city. The capacity of the city to undertake the work had been much discussed in the courts, and the Supreme Court of the State had disposed of that phase of the situation by suggesting that it did not make much difference to the municipality whether or not the debt limit permitted a contract for the work, because if the limit should be exceeded, "no liability could possibly be imposed upon the city," a view which might comfort the timid taxpayers but could hardly be expected to give confidence to the capitalists who might undertake the execution of the contract.

Various corporations, organized during the thirty odd years of unsuccessful attempts by the city to secure underground rapid transit, claimed that their franchises gave them vested rights in the streets to the exclusion of the new enterprise, and they were prepared to assert their rights in the courts. (The Underground Railroad Company of the City of New York sought to enjoin the building of the road and carried their contest to the Supreme Court of the United States which did not finally decide the questions raised until March, 1904, when the subway was practically complete.)

Rival transportation companies stood ready to obstruct the work and encourage whomever might find objection to the building of the road.

New York has biennial elections. The road could not be completed in two years, and the attitude of one administration might not be the attitude of its successors.

The engineering difficulties were well-nigh appalling. Towering buildings along the streets had to be considered, and the streets themselves were already occupied with a complicated network of subsurface structures, such as sewers, water and gas mains, electric cable conduits, electric surface railway conduits, telegraph and power conduits, and many vaults extending out under the streets, occupied by the abutting property owners. On the surface were street railway lines carrying a very heavy traffic night and day, and all the thoroughfares in the lower part of the city were congested with vehicular traffic.

Finally, the city was unwilling to take any risk, and demanded millions of dollars of security to insure the completion of the road according to the contract, the terms of which were most exacting down to the smallest detail.

The builders of the road did not underestimate the magnitude of the task before them. They retained the most experienced experts for every part of the work and, perfecting an organization in an incredibly short time, proceeded to surmount and sweep aside difficulties. The result is one of which every citizen of New York may feel proud. Upon the completion of the road the city will own the best constructed and best equipped intraurban rapid transit railroad in the world. The efforts of the builders have not been limited by the strict terms of the contract. They have striven, not to equal the best devices, but to improve upon the best devices used in modern electrical railroading, to secure for the traveling public safety, comfort, and speedy transportation.

The road is off the surface and escapes the delays incident to congested city streets, but near the surface and accessible, light, dry, clean, and well ventilated. The stations and approaches are commodious, and the stations themselves furnish conveniences to passengers heretofore not heard of on intraurban lines. There is a separate express service, with its own tracks, and the stations are so arranged that passengers may pass from local trains to express trains, and vice versa, without delay and without payment of additional fare. Special precautions have been taken and devices adopted to prevent a failure of the electric power and the consequent delays of traffic. An electro pneumatic block signal system has been devised, which excels any system heretofore used and is unique in its mechanism. The third rail for conveying the electric current is covered, so as to prevent injury to passengers and employees from contact. Special emergency and fire alarm signal systems are installed throughout the length of the road. At a few stations, where the road is not near the surface, improved escalators and elevators are provided. The cars have been designed to prevent danger from fire, and improved types of motors have been adopted, capable of supplying great speed combined with complete control. Strength, utility, and convenience have not alone been considered, but all parts of the railroad structures and equipment, stations, power house, and electrical sub-stations have been designed and constructed with a view to the beauty of their appearance, as well as to their efficiency.

The completion of the subway marks the solution of a problem which for over thirty years baffled the people of New York City, in spite of the best efforts of many of its foremost citizens. An extended account of Rapid Transit Legislation would be out of place here, but a brief glance at the history of the Act under the authority of which the subway has been built is necessary to a clear understanding of the work which has been accomplished. From 1850 to 1865 the street surface horse railways were sufficient for the requirements of the traveling public. As the city grew rapidly, the congestion spreading northward, to and beyond the Harlem River, the service of surface roads became entirely inadequate. As early as 1868, forty-two well known business men of the city became, by special legislative Act, incorporators of the New York City Central Underground Railway Company, to build a line from the City Hall to the Harlem River. The names of the incorporators evidenced the seriousness of the attempt, but nothing came of it. In 1872, also by special Act, Cornelius Vanderbilt and others were incorporated as The New York City Rapid Transit Company, to build an underground road from the City Hall to connect with the New York & Harlem Road at 59th Street, with a branch to the tracks of the New York Central Road. The enterprise was soon abandoned. Numerous companies were incorporated in the succeeding years under the general railroad laws, to build underground roads, but without results; among them the Central Tunnel Railway Company in 1881, The New York & New Jersey Tunnel Railway Company in 1883, The Terminal Underground Railway Company in 1886, The Underground Railroad Company of the City of New York (a consolidation of the last two companies) in 1896, and The Rapid Transit Underground Railroad Company in 1897.

All attempts to build a road under the early special charter and later under the general laws having failed, the city secured in 1891 the passage of the Rapid Transit Act under which, as amended, the subway has been built. As originally passed it did not provide for municipal ownership. It provided that a board of five rapid transit railroad commissioners might adopt routes and general plans for a railroad, obtain the consents of the local authorities and abutting property owners, or in lieu of the consents of the property owners the approval of the Supreme Court; and then, having adopted detail plans for the construction and operation, might sell at public sale the right to build and operate the road to a corporation, whose powers and duties were defined in the Act, for such period of time and on such terms as they could. The Commissioners prepared plans and obtained the consents of the local authorities. The property owners refused their consent; the Supreme Court gave its approval in lieu thereof, but upon inviting bids the Board of Rapid Transit Railroad Commissioners found no responsible bidder.

The late Hon. Abram S. Hewitt, as early as 1884, when legislation for underground roads was under discussion, had urged municipal ownership. Speaking in 1901, he said of his efforts in 1884:

"It was evident to me that underground rapid transit could not be secured by the investment of private capital, but in some way or other its construction was dependent upon the use of the credit of the City of New York. It was also apparent to me that if such credit were used, the property must belong to the city. Inasmuch as it would not be safe for the city to undertake the construction itself, the intervention of a contracting company appeared indispensable. To secure the city against loss, this company must necessarily be required to give a sufficient bond for the completion of the work and be willing to enter into a contract for its continued operation under a rental which would pay the interest upon the bonds issued by the city for the construction, and provide a sinking fund sufficient for the payment of the bonds at or before maturity. It also seemed to be indispensable that the leasing company should invest in the rolling stock and in the real estate required for its power houses and other buildings an amount of money sufficiently large to indemnify the city against loss in case the lessees should fail in their undertaking to build and operate the railroad."

Mr. Hewitt became Mayor of the city in 1887, and his views were presented in the form of a Bill to the Legislature in the following year. The measure found practically no support. Six years later, after the Rapid Transit Commissioners had failed under the Act of 1891, as originally drawn, to obtain bidders for the franchise, the New York Chamber of Commerce undertook to solve the problem by reverting to Mr. Hewitt's idea of municipal ownership. Whether or not municipal ownership would meet the approval of the citizens of New York could not be determined; therefore, as a preliminary step, it was decided to submit the question to a popular vote. An amendment to the Act of 1891 was drawn (Chapter 752 of the Laws of 1894) which provided that the qualified electors of the city were to decide at an annual election, by ballot, whether the rapid transit railway or railways should be constructed by the city and at the public's expense, and be operated under lease from the city, or should be constructed by a private corporation under a franchise to be sold in the manner attempted unsuccessfully, under the Act of 1891, as originally passed. At the fall election of 1894, the electors of the city, by a very large vote, declared against the sale of a franchise to a private corporation and in favor of ownership by the city. Several other amendments, the necessity for which developed as plans for the railway were worked out, were made up to and including the session of the Legislature of 1900, but the general scheme for rapid transit may be said to have become fixed when the electors declared in favor of municipal ownership. The main provisions of the legislation which stood upon the statute books as the Rapid Transit Act, when the contract was finally executed, February 21, 1900, may be briefly summarized as follows:

(*a*) The Act was general in terms, applying to all cities in the State having a population of over one million; it was special in effect because New York was the only city having such a population. It did not limit the Rapid Transit Commissioners to the building of a single road, but authorized the laying out of successive roads or extensions.

(*b*) A Board was created consisting of the Mayor, Comptroller, or other chief financial officer of the city; the president of the Chamber of Commerce of the State of New York, by virtue of his office, and five members named in the Act: William Steinway, Seth Low, John Claflin, Alexander E. Orr, and John H. Starin, men distinguished for their business experience, high integrity, and civic pride. Vacancies in the Board were to be filled by the Board itself, a guaranty of a continued uniform policy.

(*c*) The Board was to prepare general routes and plans and submit the question of municipal ownership to the electors of the city.

(*d*) The city was authorized, in the event that the electors decided for city ownership, to issue bonds not to exceed $50,000,000 for the construction of the road or roads and $5,000,000 additional, if necessary, for acquiring property rights for the route. The interest on the bonds was not to exceed 3-1/2 per cent.

(*e*) The Commissioners were given the broad power to enter into a contract (in the case of more than one road, successive contracts) on behalf of the city for the construction of the road with the person, firm, or corporation which in the opinion of the Board should be best qualified to carry out the contract, and to determine the amount of the bond to be given by the contractor to secure its performance. The essential features of the contract were, however, prescribed by the Act. The contractor in and by the contract for building the road was to agree to fully equip it at his own expense, and the equipment was to include all power houses. He was also to operate the road, as lessee of the city, for a term not to exceed fifty years, upon terms to be included in the contract for construction, which might include provision for renewals of the lease upon such terms as the Board should from time to time determine. The rental was to be at least equal to the amount of interest on the bonds which the city might issue for construction and one per cent. additional. The one per cent. additional might, in the discretion of the Board, be made contingent in part for the first ten years of the lease upon the earnings of the road. The rental was to be applied by the city to the interest on the bonds and the balance was to be paid into the city's general sinking fund for payment of the city's debt or into a sinking fund for the redemption at maturity of the bonds issued for the construction of the rapid transit road, or roads. In addition to the security which might be required by the Board of the contractor for construction and operation, the Act provided that the city should have a first lien upon the equipment of the road to be furnished by the contractor, and at the termination of the lease the city had the privilege of purchasing such equipment from the contractor.

(*f*) The city was to furnish the right of way to the contractor free from all claims of abutting property owners. The road was to be the absolute property of the city and to be deemed a part of the public streets and highways. The equipment of the road was to be exempt from taxation.

(*g*) The Board was authorized to include in the contract for construction provisions in detail for the supervision of the city, through the Board, over the operation of the road under the lease.

One of the most attractive—and, in fact, indispensable features of the scheme—was that the work of construction, instead of being subject to the conflicting control of various departments of the City Government, with their frequent changes in personnel, was under the exclusive supervision and control of the Rapid Transit Board, a conservative and continuous body composed of the two principal officers of the City Government, and five merchants of the very highest standing in the community.

Provided capitalists could be found to undertake such an extensive work under the exacting provisions, the scheme was an admirable one from the taxpayers' point of view. The road would cost the city practically nothing and the obligation of the contractor to equip and operate being combined with the agreement to construct furnished a safeguard against waste of the public funds and insured the prompt completion of the road. The interest of the contractor in the successful operation, after construction, furnished a strong incentive to see that as the construction progressed the details were consistent with successful operation and to suggest and consent to such modifications of the contract plans as might appear necessary from an operating point of view, from time to time. The rental being based upon the cost encouraged low bids, and the lien of the city upon the equipment secured the city against all risk, once the road was in operation.

Immediately after the vote of the electors upon the question of municipal ownership, the Rapid Transit Commissioners adopted routes and plans which they had been studying and perfecting since the failure to find bidders for the franchise under the original Act of 1891. The local authorities approved them, and again the property owners refused their consent, making an application to the Supreme Court necessary. The Court refused its approval upon the ground that the city, owing to a provision of the constitution of the State limiting the city's power to incur debt, would be unable to raise the necessary money. This decision appeared to nullify all the efforts of the public spirited citizens composing the Board of Rapid Transit Commissioners and to practically prohibit further attempts on their part. They persevered, however, and in January, 1897, adopted new general routes and plans. The consolidation of a large territory into the Greater New York, and increased land values, warranted the hope that the city's debt limit would no longer be an objection, especially as the new route changed the line so as to reduce the estimated cost. The demands for rapid transit had become more and more imperative as the years went by, and it was fair to assume that neither the courts nor the municipal authorities would be overzealous to find a narrow construction of the laws. Incidentally, the constitutionality of the rapid transit legislation, in its fundamental features, had been upheld in the Supreme Court in a decision which was affirmed by the highest court of the State a few weeks after the Board had adopted its new plans. The local authorities gave their consent to the new route; the property owners, as on the two previous occasions, refused their consent; the Supreme Court gave its approval in lieu thereof; and the Board was prepared to undertake the preliminaries for letting a contract. These successive steps and the preparation of the terms of the contract all took time; but, finally, on November 15, 1899, a form of contract was adopted and an invitation issued by the Board to contractors to bid for the construction and operation of the railroad. There were two bidders, one of whom was John B. McDonald, whose terms submitted under the invitation were accepted on January 15, 1900; and, for the first time, it seemed as if a beginning might be made in the actual construction of the rapid transit road. The letter of invitation to contractors required that every proposal should be accompanied by a certified check upon a National or State Bank, payable to the order of the Comptroller, for $150,000, and that within ten days after acceptance, or within such further period as might be prescribed by the Board, the contract should be duly executed and delivered. The amount to be paid by the city for the construction was $35,000,000 and an additional sum not to exceed $2,750,000 for terminals, station sites, and other purposes. The construction was to be completed in four years and a half, and the term of the lease from the city to the contractor was fixed at fifty years, with a renewal, at the option of the contractor, for twenty-five years at a rental to be agreed upon by the city, not less than the average rental for the then preceding ten years. The rental for the fifty-year term was fixed at an amount equal to the annual interest upon the bonds issued by the city for construction and 1 per cent. additional, such 1 per cent. during the first ten years to be contingent in part upon the earnings of the road. To secure the performance of the contract by Mr. McDonald the city required him to deposit $1,000,000 in cash as security for construction, to furnish a bond with surety for $5,000,000 as security for construction and equipment, and to furnish another bond of $1,000,000 as continuing security for the performance of the contract. The city in addition to this security had, under the provisions of the Rapid Transit Act, a first lien on the equipment, and it should be mentioned that at the expiration of the lease and renewals (if any) the equipment is to be turned over to the city, pending an agreement or arbitration upon the question of the price to be paid for it by the city. The contract (which covered about 200 printed pages) was minute in detail as to the work to be done, and sweeping powers of supervision were given the city through the Chief Engineer of the Board, who by the contract was made arbiter of all questions that might arise as to the interpretation of the plans and specifications. The city had been fortunate in securing for the preparation of plans the services of Mr. William Barclay Parsons, one of the foremost engineers of the country. For years as Chief Engineer of the Board he had studied and developed the various plans and it was he who was to superintend on behalf of the city the completion of the work.

During the thirty-two years of rapid transit discussion between 1868, when the New York City Central Underground Company was incorporated, up to 1900, when the invitations for bids were issued by the city, every scheme for rapid transit had failed because responsible capitalists could not be found willing to undertake the task of building a road. Each year had increased the difficulties attending such an enterprise and the scheme finally evolved had put all of the risk upon the capitalists who might attempt to finance the work, and left none upon the city. Without detracting from the credit due the public-spirited citizens who had evolved the plan of municipal ownership, it may be safely asserted that the success of the undertaking depended almost entirely upon the financial backing of the contractor. When the bid was accepted by the city no arrangements had been made for the capital necessary to carry out the contract. After its acceptance, Mr. McDonald not only found little encouragement in his efforts to secure the capital, but discovered that the surety companies were unwilling to furnish the security required of him, except on terms impossible for him to fulfill.

The crucial point in the whole problem of rapid transit with which the citizens of New York had struggled for so many years had been reached, and failure seemed inevitable. The requirements of the Rapid Transit Act were rigid and forbade any solution of the problem which committed the city to share in the risks of the undertaking. Engineers might make routes and plans, lawyers might draw legislative acts, the city might prepare contracts, the question was and always had been, Can anybody build the road who will agree to do it and hold the city safe from loss?

It was obvious when the surety companies declined the issue that the whole rapid transit problem was thrown open, or rather that it always had been open. The final analysis had not been made. After all, the attitude of the surety companies was only a reflection of the general feeling of practical business and railroad men towards the whole venture. To the companies the proposition had come as a concrete business proffer and they had rejected it.

At this critical point, Mr. McDonald sought the assistance of Mr. August Belmont. It was left to Mr. Belmont to make the final analysis, and avert the failure which impended. There was no time for indecision or delay. Whatever was to be done must be done immediately. The necessary capital must be procured, the required security must be given, and an organization for building and operating the road must be anticipated. Mr. Belmont looking through and beyond the intricacies of the Rapid Transit Act, and the complications of the contract, saw that he who undertook to surmount the difficulties presented by the attitude of the surety companies must solve the whole problem. It was not the ordinary question of financing a railroad contract. He saw that the responsibility for the entire rapid transit undertaking must be centered, and that a compact and effective organization must be planned which could deal with every phase of the situation.

Mr. Belmont without delay took the matter up directly with the Board of Rapid Transit Railroad Commissioners, and presented a plan for the incorporation of a company to procure the security required for the performance of the contract, to furnish the capital necessary to carry on the work, and to assume supervision over the whole undertaking. Application was to be made to the Supreme Court to modify the requirements with respect to the sureties by striking out a provision requiring the justification of the sureties in double the amount of liabilities assumed by each and reducing the minimum amount permitted to be taken by each surety from $500,000 to $250,000. The new corporation was to execute as surety a bond for $4,000,000, the additional amount of $1,000,000 to be furnished by other sureties. A beneficial interest in the bonds required from the sub-contractors was to be assigned to the city and, finally, the additional amount of $1,000,000, in cash or securities, was to be deposited with the city as further security for the performance of the contract. The plan was approved by the Board of Rapid Transit Railroad Commissioners, and pursuant to the plan, the Rapid Transit Subway Construction Company was organized. The Supreme Court granted the application to modify the requirements as to the justification of sureties and the contract was executed February 21, 1900.

As president and active executive head of the Rapid Transit Subway Construction Company, Mr. Belmont perfected its organization, collected the staff of engineers under whose direction the work of building the road was to be done, supervised the letting of sub-contracts, and completed the financial arrangements for carrying on the work.

The equipment of the road included, under the terms of the contract, the rolling stock, all machinery and mechanisms for generating electricity for motive power, lighting, and signaling, and also the power house, sub-stations, and the real estate upon which they were to be erected. The magnitude of the task of providing the equipment was not generally appreciated until Mr. Belmont took the rapid transit problem in hand. He foresaw from the beginning the importance of that branch of the work, and early in 1900, immediately after the signing of the contract, turned his attention to selecting the best engineers and operating experts, and planned the organization of an operating company. As early as May, 1900, he secured the services of Mr. E. P. Bryan, who came to New York from St. Louis, resigning as vice-president and general manager of the Terminal Railroad Association, and began a study of the construction work and plans for equipment, to the end that the problems of operation might be anticipated as the building and equipment of the road progressed. Upon the incorporation of the operating company, Mr. Bryan became vice-president.

In the spring of 1902, the Interborough Rapid Transit Company, the operating railroad corporation was formed by the interests represented by Mr. Belmont, he becoming president and active executive head of this company also, and soon thereafter Mr. McDonald assigned to it the lease or operating part of his contract with the city, that company thereby becoming directly responsible to the city for the equipment and operation of the road, Mr. McDonald remaining as contractor for its construction. In the summer of the same year, the Board of Rapid Transit Railroad Commissioners having adopted a route and plans for an extension of the subway under the East River to the Borough of Brooklyn, the Rapid Transit Subway Construction Company entered into a contract with the city, similar in form to Mr. McDonald's contract, to build, equip, and operate the extension. Mr. McDonald, as contractor of the Rapid Transit Subway Construction Company, assumed the general supervision of the work of constructing the Brooklyn extension; and the construction work of both the original subway and the extension has been carried on under his direction. The work of construction has been greatly facilitated by the broad minded and liberal policy of the Rapid Transit Board and its Chief Engineer and Counsel, and by the coöperation of all the other departments of the City Government, and also by the generous attitude of the Metropolitan Street Railway Company and its lessee, the New York City Railroad Company, in extending privileges which have been of great assistance in the prosecution of the work. In January, 1903, the Interborough Rapid Transit Company acquired the elevated railway system by lease for 999 years from the Manhattan Railway Company, thus assuring harmonious operation of the elevated roads and the subway system, including the Brooklyn extension.

The incorporators of the Interborough Rapid Transit Company were William H. Baldwin, Jr., Charles T. Barney, August Belmont, E. P. Bryan, Andrew Freedman, James Jourdan, Gardiner M. Lane, John B. McDonald, DeLancey Nicoll, Walter G. Oakman, John Peirce, Wm. A. Read, Cornelius Vanderbilt, George W. Wickersham, and George W. Young.

The incorporators of the Rapid Transit Subway Construction Company were Charles T. Barney, August Belmont, John B. McDonald, Walter G. Oakman, and William A. Read.

EXTERIOR VIEW OF POWER HOUSE

CHAPTER I

THE ROUTE OF THE ROAD—
PASSENGER STATIONS AND TRACKS

The selection of route for the Subway was governed largely by the amount which the city was authorized by the Rapid Transit Act to spend. The main object of the road was to carry to and from their homes in the upper portions of Manhattan Island the great army of workers who spend the business day in the offices, shops, and warehouses of the lower portions, and it was therefore obvious that the general direction of the routes must be north and south, and that the line must extend as nearly as possible from one end of the island to the other.

The routes proposed by the Rapid Transit Board in 1895, after municipal ownership had been approved by the voters at the fall election of 1894, contemplated the occupation of Broadway below 34th Street to the Battery, and extended only to 185th Street on the west side and 146th Street on the east side of the city. As has been told in the introductory chapter, this plan was rejected by the Supreme Court because of the probable cost of going under Broadway. It was also intimated by the Court, in rejecting the routes, that the road should extend further north.

It had been clear from the beginning that no routes could be laid out to which abutting property owners would consent, and that the consent of the Court as an alternative would be necessary to any routes chosen. To conform as nearly as possible to the views of the Court, the Commission proposed, in 1897, the so called "Elm Street route," the plan finally adopted, which reached from the territory near the General Post-office, the City Hall, and Brooklyn Bridge Terminal to Kingsbridge and the station of the New York & Putnam Railroad on the upper west side, and to Bronx Park on the upper east side of the city, touching the Grand Central Depot at 42d Street.

Subsequently, by the adoption of the Brooklyn Extension, the line was extended down Broadway to the southern extremity of Manhattan Island, thence under the East River to Brooklyn.

The routes in detail are as follows:

Manhattan-Bronx Route

Beginning near the intersection of Broadway and Park Row, one of the routes of the railroad extends under Park Row, Center Street, New Elm Street, Elm Street, Lafayette Place, Fourth Avenue (beginning at Astor Place), Park Avenue, 42d Street, and Broadway to 125th Street, where it passes over Broadway by viaduct to 133d Street, thence under Broadway again to and under Eleventh Avenue to Fort George, where it comes to the surface again at Dyckman Street and continues by viaduct over Naegle Avenue, Amsterdam Avenue, and Broadway to Bailey Avenue, at the Kingsbridge station of the New York & Putnam Railroad, crossing the Harlem Ship Canal on a double-deck drawbridge. The length of this route is 13.50 miles, of which about 2 miles are on viaduct.

Another route begins at Broadway near 103d Street and extends under 104th Street and the upper part of Central Park to and under Lenox Avenue to 142d Street, thence curving to the east to and under the Harlem River at about 145th Street, thence from the river to and under East 149th Street to a point near Third Avenue, thence by viaduct beginning at Brook Avenue over Westchester Avenue, the Southern Boulevard and the Boston Road to Bronx Park. The length of this route is about 6.97 miles, of which about 3 miles are on viaduct.

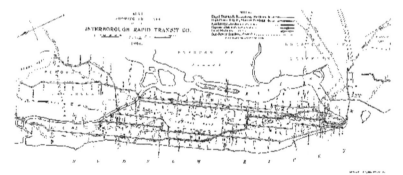

MAP SHOWING THE LINES OF THE INTERBOROUGH RAPID TRANSIT CO. 1904

At the City Hall there is a loop under the Park. From 142d Street there is a spur north under Lenox Avenue to 148th Street. There is a spur at Westchester and Third Avenues connecting by viaduct the Manhattan Elevated Railway Division of Interborough Rapid Transit Company with the viaduct of the subway at or near St. Ann's Avenue.

Brooklyn Route

The route of the Brooklyn Extension connects near Broadway and Park Row with the Manhattan Bronx Route and extends under Broadway,

Bowling Green, State Street, Battery Park, Whitehall Street, and South Street to and under the East River to Brooklyn at the foot of Joralemon Street, thence under Joralemon Street, Fulton Street, and Flatbush Avenue to Atlantic Avenue, connecting with the Brooklyn tunnel of the Long Island Railroad at that point. There is a loop under Battery Park beginning at Bridge Street. The length of this route is about 3 miles.

The routes in Manhattan and The Bronx may therefore be said to roughly resemble the letter Y with the base at the southern extremity of Manhattan Island, the fork at 103d Street and Broadway, the terminus of the westerly or Fort George branch of the fork just beyond Spuyten Duyvil Creek, the terminus of the easterly or Bronx Park branch at Bronx Park.

Location of Stations

The stations beginning at the base of the Y and following the route up to the fork are located at the following points:

South Ferry, Bowling Green and Battery Place, Rector Street and Broadway, Fulton Street and Broadway, City Hall, Manhattan; Brooklyn Bridge Entrance, Manhattan; Worth and Elm Streets, Canal and Elm Streets, Spring and Elm Streets, Bleecker and Elm Streets, Astor Place and Fourth Avenue, 14th Street and Fourth Avenue, 18th Street and Fourth Avenue, 23d Street and Fourth Avenue, 28th Street and Fourth Avenue, 33d Street and Fourth Avenue, 42d Street and Madison Avenue (Grand Central Station), 42d Street and Broadway, 50th Street and Broadway, 60th Street and Broadway (Columbus Circle), 66th Street and Broadway, 72d Street and Broadway, 79th Street and Broadway, 86th Street and Broadway, 91st Street and Broadway, 96th Street and Broadway.

34TH STREET AND PARK AVENUE, LOOKING SOUTH

The stations of the Fort George or westerly branch are located at the following points:

One Hundred and Third Street and Broadway, 110th Street and Broadway (Cathedral Parkway), 116th Street and Broadway (Columbia University), Manhattan Street (near 128th Street) and Broadway, 137th Street and Broadway, 145th Street and Broadway, 157th Street and Broadway, the intersection of 168th Street, St. Nicholas Avenue and Broadway, 181st Street and Eleventh Avenue, Dyckman Street and Naegle Avenue (beyond Fort George), 207th Street and Amsterdam Avenue, 215th Street and Amsterdam Avenue, Muscoota Street and Broadway, Bailey Avenue, at Kingsbridge near the New York & Putnam Railroad station.

The stations on the Bronx Park or easterly branch are located at the following points:

One Hundred and Tenth Street and Lenox Avenue, 116th Street and Lenox Avenue, 125th Street and Lenox Avenue, 135th Street and Lenox Avenue, 145th Street and Lenox Avenue (spur), Mott Avenue and 149th Street, the intersection of 149th Street, Melrose and Third Avenues, Jackson and Westchester Avenues, Prospect and Westchester Avenues, Westchester Avenue near Southern Boulevard (Fox Street), Freeman Street and the Southern Boulevard, intersection of 174th Street, Southern Boulevard and Boston Road, 177th Street and Boston Road (near Bronx Park).

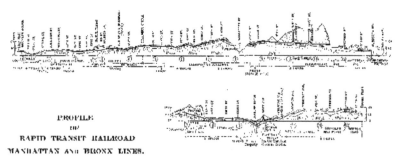

PROFILE
OF
RAPID TRANSIT RAILROAD
MANHATTAN AND BRONX LINES.

PROFILE OF RAPID TRANSIT RAILROAD
MANHATTAN AND BRONX LINES.

The stations in the Borough of Brooklyn on the Brooklyn Extension are located as follows:

Joralemon Street near Court (Brooklyn Borough Hall), intersection of Fulton, Bridge, and Hoyt Streets; Flatbush Avenue near Nevins Street, Atlantic Avenue and Flatbush Avenue (Brooklyn terminal of the Long Island Railroad).

From the Borough Hall, Manhattan, to the 96th Street station, the line is four-track. On the Fort George branch (including 103d Street station) there are three tracks to 145th Street and then two tracks to Dyckman Street, then three tracks again to the terminus at Bailey Avenue. On the Bronx Park branch there are two tracks to Brook Avenue and from that point to Bronx Park there are three tracks. On the Lenox Avenue spur to 148th Street there are two tracks, on the City Hall loop one track, on the Battery Park loop two tracks. The Brooklyn Extension is a two-track line.

There is a storage yard under Broadway between 137th Street and 145th Street on the Fort George branch, another on the surface at the end of the Lenox Avenue spur, Lenox Avenue and 148th Street, and a third on an elevated structure at the Boston Road and 178th Street. There is a repair shop and inspection shed on the surface adjoining the Lenox Avenue spur at the Harlem River and 148-150th Streets, and an inspection shed at the storage yard at Boston Road and 178th Street.

Length of Line.

The total length of the line from the City Hall to the Kingsbridge terminal is 13.50 miles, with 47.11 miles of single track and sidings. The eastern or Bronx Park branch is 6.97 miles long, with 17.50 miles of single track.

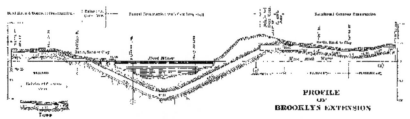

PROFILE OF BROOKLYN EXTENSION.

Grades and Curves.

The total length of the Brooklyn Extension is 3.1 miles, with about 8 miles of single track.

The grades and curvature along the main line may be summarized as follows:

The total curvature is equal in length to 23 per cent. of the straight line, and the least radius of curvature is 147 feet. The greatest grade is 3 per cent., and occurs on either side of the tunnel under the Harlem River. At each station there is a down grade of 2.1 per cent., to assist in the acceleration of the cars when they start. In order to make time on roads running trains at frequent intervals, it is necessary to bring the trains to their full speed very soon after starting. The electrical equipment of the Rapid Transit Railroad will enable this to be done in a better manner than is possible with steam locomotives, while these short acceleration grades at each station, on both up and down tracks, will be of material assistance in making the starts smooth.

Photograph on page 26 shows an interesting feature at a local station, where, in order to obtain the quick acceleration in grade for local trains, and at the same time maintain a level grade for the express service, the tracks are constructed at a different level. This occurs at many local stations.

On the Brooklyn Extension the maximum grade is 3.1 per cent. descending from the ends to the center of the East River tunnel. The minimum radius of curve is 1,200 feet.

**STANDARD STEEL CONSTRUCTION IN TUNNEL—THIRD
RAIL PROTECTION NOT SHOWN**

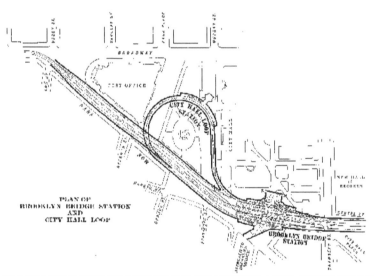

**PLAN OF BROOKLYN BRIDGE STATION AND CITY HALL
LOOP**

Track

The track is of the usual standard construction with broken stone ballast, timber cross ties, and 100-pound rails of the American Society of Civil Engineers' section. The cross ties are selected hard pine. All ties are fitted with tie plates. All curves are supplied with steel inside guard rails. The frogs and switches are of the best design and quality to be had, and a special design has been used on all curves. At the Battery loop, at Westchester Avenue, at 96th Street, and at City Hall loop, where it has been necessary for the regular passenger tracks to cross, grade crossings have been avoided; one track or set of tracks passing under the other at the intersecting points. (See plan on this page.)

The contract for the building of the road contains the following somewhat unusual provision: "The railway and its equipment as contemplated by the contract constitute a great public work. All parts of the structure where exposed to public sight shall therefore be designed, constructed, and maintained with a view to the beauty of their appearance, as well as to their efficiency."

It may be said with exact truthfulness that the builders have spared no effort or expense to live up to the spirit of this provision, and that all parts of the road and equipment display dignified and consistent artistic effects of the highest order. These are noticeable in the power house and the electrical sub-stations and particularly in the passenger stations. It might readily have been supposed that the limited space and comparative uniformity of the underground stations would afford but little opportunity for architectural and decorative effects. The result has shown the fallacy of such a supposition.

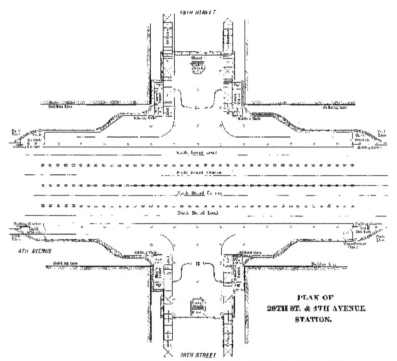

PLAN OF 28TH ST. & 4TH AVENUE STATION.

Of the forty-eight stations, thirty-three are underground, eleven are on the viaduct portions of the road, and three are partly on the surface and partly underground, and one is partly on the surface and partly on the viaduct.

Space Occupied

The underground stations are at the street intersections, and, except in a few instances, occupy space under the cross streets. The station plans are necessarily varied to suit the conditions of the different locations, the most important factor in planning them having been the amount of available space. The platforms are from 200 to 350 feet in length, and about 16 feet in width, narrowing at the ends, while the center space is larger or smaller, according to local conditions. As a rule the body of the station extends back about 50 feet from the edge of the platform.

At all local stations (except at 110th Street and Lenox Avenue) platforms are outside of the tracks. (Plan and photograph on pages 30 and 31.) At Lenox Avenue and 110th Street there is a single island platform for uptown and downtown passengers.

At express stations there are two island platforms between the express and local tracks, one for uptown and one for downtown traffic. In addition, there are the usual local platforms at Brooklyn Bridge, 14th Street (photograph on page 34) and 96th Street. At the remaining express stations, 42d Street and Madison Avenue and 72d Street, there are no local platforms outside of the tracks, local and through traffic using the island platforms.

28TH STREET STATION

The island platforms at Brooklyn Bridge, 14th Street, and 42d Street and Madison Avenue are reached by mezzanine footways from the local platforms, it having been impossible to place entrances in the streets immediately over the platforms. At 96th Street there is an underground passage connecting the local and island platforms, and at 72d Street there are entrances to the island platforms directly from the street because there is a park area in the middle of the street. Local passengers can transfer from express trains and express passengers from local trains without payment of additional fare by stepping across the island platforms.

At 72d Street, at 103d Street, and at 116th Street and Broadway the station platforms are below the surface, but the ticket booths and toilet rooms are on the surface; this arrangement being possible also because of the park area available in the streets. At Manhattan Street the platforms are on the viaduct, but the ticket booths and toilet rooms are on the surface. The viaduct at this point is about 68 feet above the surface, and escalators are provided. At many of the stations entrances have been arranged from the adjacent buildings, in addition to the entrances originally planned from the street.

Kiosks

The entrances to the underground stations are enclosed at the street by kiosks of cast iron and wire glass (photograph on page 33), and vary in number from two to eight at a station. The stairways are of concrete, reinforced by twisted steel rods. At 168th Street, at 181st Street, and at Mott Avenue, where the platforms are from 90 to 100 feet below the surface, elevators are provided.

WEST SIDE
OF 23D STREET STATION

At twenty of the underground stations it has been possible to use vault lights to such an extent that very little artificial light is needed. (Photograph on page 35.) Such artificial light as is required is supplied by incandescent lamps sunk in the ceilings. Provision has been made for using the track circuit for lighting in emergency if the regular lighting circuit should temporarily fail.

KIOSKS AT COLUMBUS CIRCLE

The station floors are of concrete, marked off in squares. At the junction of the floors and side walls a cement sanitary cove is placed. The floors drain to catch-basins, and hose bibs are provided for washing the floors.

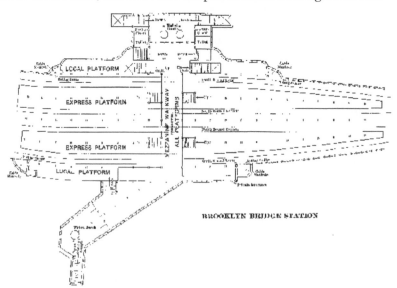

BROOKLYN BRIDGE STATION

Two types of ceiling are used, one flat, which covers the steel and concrete of the roof, and the other arched between the roof beams and girders, the lower flanges of which are exposed. Both types have an air space between ceiling and roof, which, together with the air space behind the inner side walls, permits air to circulate and minimizes condensation on the surface of the ceiling and walls.

PLAQUE SHOWING BEAVER AT ASTOR PLACE STATION

The ceilings are separated into panels by wide ornamental mouldings, and the panels are decorated with narrower mouldings and rosettes. The bases of the walls are buff Norman brick. Above this is glass tile or glazed tile, and above the tile is a faience or terra-cotta cornice. Ceramic mosaic is used for decorative panels, friezes, pilasters, and name-tablets. A different decorative treatment is used at each station, including a distinctive color scheme. At some stations the number of the intersecting street or initial letter of the street name is shown on conspicuous plaques, at other stations the number or letter is in the panel. At some stations artistic emblems have been used in the scheme of decoration, as at Astor Place, the beaver (see photograph on this page); at Columbus Circle, the great navigator's Caravel; at 116th Street, the seal of Columbia University. The walls above the cornice and the ceilings are finished in white Keene cement.

**EXPRESS STATION AT 14TH STREET, SHOWING ISLAND
AND MEZZANINE PLATFORMS AND STAIRS CONNECTING
THEM**

**WEST SIDE OF COLUMBUS CIRCLE STATION (60TH
STREET)—ILLUMINATED BY DAYLIGHT COMING
THROUGH VAULT LIGHTS**

CARAVEL AND WALL DECORATION

The ticket booths are of oak with bronze window grills and fittings. There are toilet rooms in every station, except at the City Hall loop. Each toilet room has a free closet or closets, and a pay closet which is furnished with a basin, mirror, soap dish, and towel rack. The fixtures are porcelain, finished in dull nickel. The soil, vent and water pipes are run in wall spaces, so as to be accessible. The rooms are ventilated through the hollow columns of the kiosks, and each is provided with an electric fan. They are heated by electric heaters. The woodwork of the rooms is oak; the walls are red slate wainscot and Keene cement.

Passengers may enter the body of the station without paying fare. The train platforms are separated from the body of the station by railings. At the more important stations, separate sets of entrances are provided for incoming and outgoing passengers, the stairs at the back of the station being used for entrances and those nearer the track being used for exits.

CITY HALL STATION

An example of the care used to obtain artistic effects can be seen at the City Hall station. The road at this point is through an arched tunnel. In order to secure consistency in treatment the roof of the station is continued by a larger arch of special design. (See photograph on this page.) At 168th Street, and at 181st Street, and at Mott Avenue stations, where the road is far beneath the surface, it has been possible to build massive arches over the stations and tracks, with spans of 50 feet.

CHAPTER II

TYPES AND METHODS OF CONSTRUCTION

Five types of construction have been employed in building the road: (1) the typical subway near the surface with flat roof and "I" beams for the roof and sides, supported between tracks with steel bulb-angle columns used on about 10.6 miles or 52.2 per cent. of the road; (2) flat roof typical subway of reënforced concrete construction supported between the tracks by steel bulb-angle columns, used for a short distance on Lenox Avenue and on the Brooklyn portion of the Brooklyn Extension, also on the Battery Park loop; (3) concrete lined tunnel used on about 4.6 miles or 23 per cent. of the road, of which 4.2 per cent. was concrete lined open cut work, and the remainder was rock tunnel work; (4) elevated road on steel viaduct used on about 5 miles or 24.6 per cent. of the road; (5) cast-iron tubes used under the Harlem and East Rivers.

Typical Subway

The general character of the flat roof "I" beam construction is shown in photograph on page 28 and drawing on this page]. The bottom is of concrete. The side walls have "I" beam columns five feet apart, between which are vertical concrete arches, the steel acting as a support for the masonry and allowing the thickness of the walls to be materially reduced from that necessary were nothing but concrete used. The tops of the wall columns are connected by roof beams which are supported by rows of steel columns between the tracks, built on concrete and cut stone bases forming part of the floor system. Concrete arches between the roof beams complete the top of the subway. Such a structure is not impervious, and hence, there has been laid behind the side walls, under the floor and over the roof a course of two to eight thicknesses of felt, each washed with hot asphalt as laid. In addition to this precaution against dampness, in three sections of the subway (viz.: on Elm Street between Pearl and Grand Streets, and on the approaches to the Harlem River tunnel, and on the Battery Park Loop) the felt waterproofing has been made more effective by one or two courses of hard-burned brick laid in hot asphalt, after the manner sometimes employed in constructing the linings of reservoirs of waterworks.

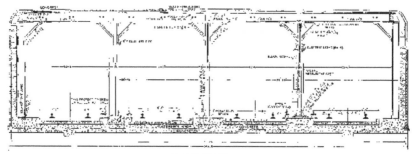

TYPICAL SECTION OF
FOUR TRACK SUBWAY

TYPICAL SECTION OF FOUR TRACK SUBWAY

FOUR-TRACK SUBWAY—SHOWING CROSS-OVER SOUTH OF 18TH STREET STATION

In front of the waterproofing, immediately behind the steel columns, are the systems of terra-cotta ducts in which the electric cables are placed. The cables can be reached by means of manholes every 200 to 450 feet, which open into the subway and also into the street. The number of these ducts ranges from 128 down to 32, and they are connected with the main power station at 58th and 59th Streets and the Hudson River by a 128-duct subway under the former street.

Reinforced Concrete Construction

The reinforced concrete construction substitutes for the steel roof beams, steel rods, approximating 1-1/4 inches square, laid in varying distances according to the different roof loads, from six to ten inches apart. Rods 1-1/8 inches in diameter tie the side walls, passing through angle columns in the walls and the bulb-angle columns in the center. Layers of concrete are laid over the roof rods to a thickness of from eighteen to thirty inches, and carried two inches below the rods, imbedding them. For the sides similar square rods and concrete are used and angle columns five feet apart. The concrete of the side walls is from fifteen to eighteen inches thick. This type is shown by photographs on page 41. The rods used are of both square and twisted form.

LAYING SHEET WATERPROOFING IN BOTTOM

SPECIAL BRICK AND ASPHALT WATERPROOFING

The construction of the typical subway has been carried on by a great variety of methods, partly adopted on account of the conditions under which the work had to be prosecuted and partly due to the personal views of the different sub-contractors. The work was all done by open excavation, the so-called "cut and cover" system, but the conditions varied widely along different parts of the line, and different means were adopted to overcome local difficulties. The distance of the rock surface below the street level had a marked influence on the manner in which the excavation of the open trenches could be made. In some places this rock rose nearly to the pavement, as between 14th and 18th Streets. At other places the subway is located in water-bearing loam and sand, as in the stretch between Pearl and Grand Streets, where it was necessary to employ a special design for the bottom, which is illustrated by drawing on page 42.

This part of the route includes the former site of the ancient Collect Pond, familiar in the early history of New York, and the excavation was through made ground, the pond having been filled in for building purposes after it was abandoned for supplying water to the city. The excavations through Canal Street, adjacent, were also through made ground, that street having been at one time, as its name implies, a canal.

From the City Hall to 9th Street was sand, presenting no particular difficulties except through the territory just described.

At Union Square rock was encountered on the west side of Fourth Avenue from the surface down. On the east side of the street, however, at the surface was sand, which extended 15 feet down to a sloping rock surface. The tendency of the sand to a slide off into the rock excavation required great care. The work was done, however, without interference with the street traffic, which is particularly heavy at that point.

**DUCTS IN SIDE WALLS—EIGHT ONLY OF THE SIXTEEN
LAYERS ARE SHOWN**

REINFORCED CONCRETE CONSTRUCTION

**ROOF SHOWING CONCRETE-STEEL CONSTRUCTION—
LENOX AVENUE AND 140TH-141ST STREETS**

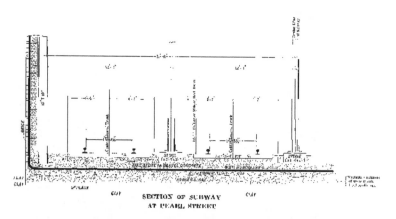

SECTION OF SUBWAY AT PEARL STREET
This construction was made necessary by encountering a layer of
Peat resting on Clay

**SURFACE RAILWAY TRACKS SUPPORTED OVER
EXCAVATION ON UPPER BROADWAY**

**SUBDIVISION OF 36" AND 30" GAS MAINS OVER ROOF OF
SUBWAY—66TH STREET AND BROADWAY**

The natural difficulties of the route were increased by the network of
sewers, water and gas mains, steam pipes, pneumatic tubes, electric
conduits and their accessories, which filled the streets; and by the surface
railways and their conduits. In some places the columns of the elevated
railway had to be shored up temporarily, and in other places the subway
passes close to the foundations of lofty buildings, where the construction
needed to insure the safety of both subway and buildings was quite
intricate. As the subway is close to the surface along a considerable part of
its route, its construction involved the reconstruction of all the

underground pipes and ducts in many places, as well as the removal of projecting vaults and buildings, and, in some cases, the underpinning of their walls. A description in detail of the methods of construction followed all along the line would make an interesting book of itself. Space will only permit, however, an account of how some of the more serious difficulties were overcome.

On Fourth Avenue, north of Union Square to 33d Street, there were two electric conduit railway tracks in the center of the roadway and a horse car track near each curb part of the distance. The two electric car tracks were used for traffic which could not be interrupted, although the horse car tracks could be removed without inconvenience. These conditions rendered it impracticable to disturb the center of the roadway, while permitting excavation near the curb. Well-timbered shafts about 8 x 10 feet, in plan, were sunk along one curb line and tunnels driven from them toward the other side of the street, stopping about 3-1/2 feet beyond its center line. A bed of concrete was laid on the bottom of each tunnel, and, when it had set, a heavy vertical trestle was built on it. In this way trestles were built half across the street, strong enough to carry all the street cars and traffic on that half of the roadway. Cableways to handle the dirt were erected near the curb line, spanning a number of these trestles, and then the earth between them was excavated from the curb to within a few feet of the nearest electric car track. The horse car tracks were removed. Between the electric tracks a trench was dug until its bottom was level with the tops of the trestles, about three feet below the surface as a rule. A pair of heavy steel beams was then laid in this trench on the trestles. Between these beams and the curb line a second pair of beams were placed. In this way the equivalent of a bridge was put up, the trestles acting as piers and the beams as girders. The central portion of the roadway was then undermined and supported by timbering suspended from the steel beams. The various gas and water pipes were hung from timbers at the surface of the ground. About four sections, or 150 feet, of the subway were built at a time in this manner. When the work was completed along one side of the street it was repeated in the same manner on the other side. This method of construction was subsequently modified so as to permit work on both sides of the street simultaneously. The manner in which the central part of the roadway was supported remained the same and all of the traffic was diverted to this strip.

SUPPORT OF ELEVATED RAILWAY STATION AT 42D STREET AND SIXTH AVENUE

Between 14th and 17th Streets, because of the proximity of the rock to the surface, it was necessary to move the tracks of the electric surface railway from the center of the street some twenty feet to the east curb, without interrupting traffic, which was very heavy at all times, the line being one of the main arteries of the Metropolitan system. Four 12 x 12-inch timbers were laid upon the surface. Standard cast-iron yokes were placed upon the timbers at the usual distance apart. Upon this structure the regular track and slot rails were placed. The space between the rails was floored over. Wooden boxes were temporarily laid for the electric cables. The usual hand holes and other accessories were built and the road operated on this timber roadbed. The removal of the tracks was made necessary because the rock beneath them and the concrete around the yokes was so closely united as to be practically monolithic, precluding the use of explosives. Attempts to remove the rock from under the track demonstrated that it could not be done without destroying the yokes of the surface railway.

SUPPORTING ELEVATED RAILROAD BY EXTENSION GIRDER—64TH STREET AND BROADWAY

The method of undermining the tracks on Broadway from 60th to 104th Streets was entirely different, for the conditions were not the same. The street is a wide one with a 22-foot parkway in the center, an electric conduit railway on either side, and outside each track a wide roadway. The subway excavation extended about 10 feet outside each track, leaving between it and the curb ample room for vehicles. The construction problem, therefore, was to care for the car tracks with a minimum interference with the excavation. This was accomplished by temporary bridges for each track, each bridge consisting of a pair of timber trusses about 55 feet long, braced together overhead high enough to let a car pass below the bracing. These trusses were set up on crib-work supports at each end, and the track hung from the lower chords. (See photograph on page 42.) The excavation then proceeded until the trench was finished and posts could be put into place between its bottom and the track. When the track was securely supported in this way, the trusses were lifted on flat cars and moved ahead 50 feet.

At 66th Street station the subway roof was about 2 feet from the electric railway yokes and structures of the street surface line. In order to build at this point it was necessary to remove two large gas mains, one 30 inches and the other 36 inches in diameter, and substitute for them, in troughs built between the roof beams of the subway, five smaller gas mains, each 24 inches in diameter. This was done without interrupting the use of the mains.

MOVING BRICK AND CONCRETE RETAINING WALL TO MAKE ROOM FOR THIRD TRACK—BROADWAY AND 134TH STREET

At the station on 42d Street, between Park and Madison Avenues, where there are five subway tracks, and along 42d Street to Broadway, a special method of construction was employed which was not followed elsewhere. The excavation here was about 35 feet deep and extended 10 to 15 feet into rock. A trench 30 feet wide was first sunk on the south side of the street and the subway built in it for a width of two tracks. Then, at intervals of 50 feet, tunnels were driven toward the north side of the street. Their tops were about 4 feet above the roof of the subway and their bottoms were on the roof. When they had been driven just beyond the line of the fourth track, their ends were connected by a tunnel parallel with the axis of the subway. The rock in the bottom of all these tunnels was then excavated to its final depth. In the small tunnel parallel with the subway axis, a bed of concrete was placed and the third row of steel columns was erected ready to carry the steel and concrete roof. When this work was completed, the earth between the traverse tunnels was excavated, the material above being supported on poling boards and struts. The roof of the subway was then extended sidewise over the rock below from the second to the third row of columns, and it was not until the roof was finished that the rock beneath was excavated. In this way the subway was finished for a width of four tracks. For the fifth track the earth was removed by tunneling to the limits of the subway, and then the rock below was blasted out.

MOVING WEST SIDE WALL TO WIDEN SUBWAY FOR
THIRD TRACK—135TH STREET AND BROADWAY

SUBWAY THROUGH NEW "TIMES" BUILDING, SHOWING
INDEPENDENT CONSTRUCTION—THE WORKMEN STAND
ON FLOOR GIRDERS OF SUBWAY

COLUMNS OF HOTEL BELMONT, PASSING THROUGH SUBWAY AT 42D STREET AND PARK AVENUE

In a number of places it was necessary to underpin the columns of the elevated railways, and a variety of methods were adopted for the work. A typical example of the difficulties involved was afforded at the Manhattan Railway Elevated Station at Sixth Avenue and 42d Street. The stairways of this station were directly over the open excavation for the subway in the latter thoroughfare and were used by a large number of people. The work was done in the same manner at each of the four corners. Two narrow pits about 40 feet apart, were first sunk and their bottoms covered with concrete at the elevation of the floor of the subway. A trestle was built in each pit, and on these were placed a pair of 3-foot plate girders, one on each side of the elevated column, which was midway between the trestles. The column was then riveted to the girders and was thus held independent of its original foundations. Other pits were then sunk under the stairway and trestles built in them to support it. When this work was completed it was possible to carry out the remaining excavation without interfering with the elevated railway traffic.

At 64th Street and Broadway, also, the whole elevated railway had to be supported during construction. A temporary wooden bent was used to carry the elevated structure. The elevated columns were removed until the subway structure was completed at that point. (See photograph on page 45.)

SMALL WATER MAINS BETWEEN STREET SURFACE AND SUBWAY ROOF, SUBSTITUTED FOR ONE LARGE MAIN— 125TH STREET AND LENOX AVE.

SPECIAL CONSTRUCTION OF 6-1/2-FOOT SEWER, UNDER CHATHAM SQUARE

A feature of the construction which attracted considerable public attention while it was in progress, was the underpinning of a part of the

Columbus Monument near the southwest entrance to Central Park. This handsome memorial column has a stone shaft rising about 75 feet above the street level and weighs about 700 tons. The rubble masonry foundation is 45 feet square and rests on a 2-foot course of concrete. The subway passes under its east side within 3 feet of its center, thus cutting out about three-tenths of the original support. At this place the footing was on dry sand of considerable depth, but on the other side of the monument rock rose within 3 feet of the surface. The steep slope of the rock surface toward the subway necessitated particular care in underpinning the footings. The work was done by first driving a tunnel 6 feet wide and 7 feet high under the monument just outside the wall line of the subway. The tunnel was given a 2-foot bottom of concrete as a support for a row of wood posts a foot square, which were put in every 5 feet to carry the footing above. When these posts were securely wedged in place the tunnel was filled with rubble masonry. This wall was strong enough to carry the weight of the portion of the monument over the subway, but the monument had to be supported to prevent its breaking off when undermined. To support it thus a small tunnel was driven through the rubble masonry foundation just below the street level and a pair of plate girders run through it. A trestle bent was then built under each end of the girders in the finished excavation for the subway. The girders were wedged up against the top of the tunnel in the masonry and the excavation was carried out under the monument without any injury to that structure.

**THREE PIPES SUBSTITUTED FOR LARGE BRICK SEWER AT
110TH STREET AND LENOX AVENUE**

SEWER SIPHON AT 149TH STREET AND RAILROAD AVENUE

**CONCRETE SEWER BACK OF ELECTRIC DUCT
MANHOLE—BROADWAY AND 58TH STREET**

At 134th Street and Broadway a two-track structure of the steel beam type about 200 feet long was completed. Approaching it from the south, leading from Manhattan Valley Viaduct, was an open cut with retaining walls 300 feet long and from 3 to 13 feet in height. After all this work was finished (and it happened to be the first finished on the subway), it was decided to widen the road to three tracks, and a unique piece of work was successfully accomplished. The retaining walls were moved bodily on slides, by means of jacks, to a line 6-1/4 feet on each side, widening the roadbed 12-1/2 feet, without a break in either wall. The method of widening the steel-beam typical subway portion was equally novel. The west wall was moved bodily by jacks the necessary distance to bring it in line with the new position of the west retaining wall. The remainder of the structure was then moved bodily, also by jacks, 6-1/4 feet to the east. The new roof of the usual type was then added over 12-1/2 feet of additional opening. (See photographs on pages 46 and 47.)

CONCRETE SEWER BACK OF SIDE WALL, BROADWAY AND 56TH STREET

**LARGE GAS AND WATER PIPES, RELAID BEHIND EACH
SIDE WALL ON ELM STREET**

Provision had to be made, not only for buildings along the route that towered far above the street surface, but also for some which burrowed far below the subway. Photograph on page 47 shows an interesting example at 42d Street and Broadway, where the pressroom of the new building of the "New York Times" is beneath the subway, the first floor is above it, and the first basement is alongside of it. Incidentally it should be noted that the steel structure of the building and the subway are independent, the columns of the building passing through the subway station.

DIFFICULT PIPE WORK—BROADWAY AND 70TH STREET

At 42d Street and Park Avenue the road passes under the Hotel Belmont, which necessitated the use of extra heavy steel girders and foundations for the support of the hotel and reinforced subway station. (See photograph on page 48.)

Along the east side of Park Row the ascending line of the "loop" was built through the pressroom of the "New York Times" (the older downtown building), and as the excavation was considerably below the bottom of the foundation of the building, great care was necessary to avoid any settlement. Instead of wood sheathing, steel channels were driven and thoroughly braced, and construction proceeded without disturbance of the building, which is very tall.

At 125th Street and Lenox Avenue one of the most complicated network of subsurface structures was encountered. Street surface electric lines with their conduits intersect. On the south side of 125th Street were a 48-inch water main and a 6-inch water main, a 12-inch and two 10-inch gas pipes and a bank of electric light and power ducts. On the north side were a 20-inch water main, one 6-inch, one 10-inch, and one 12-inch gas pipe and two banks of electric ducts. The headroom between the subway roof and the surface of the street was 4.75 feet. It was necessary to relocate the yokes of the street railway tracks on Lenox Avenue so as to bring them directly over the tunnel roof-beams. Between the lower flanges of the roof-beams, for four bents, were laid heavy steel plates well stiffened, and in these troughs were laid four 20-inch pipes, which carried the water of the 48-inch main. (See photograph on page 49.) Special castings were necessary to make the connections at each end. The smaller pipes and ducts were rearranged and carried over the roof or laid in troughs composed of 3-inch I-beams laid on the lower flanges of the roof-beams. In addition to all the transverse pipes, there were numerous pipes and duct lines to be relaid and rebuilt parallel to the subway and around the station. The change was accomplished without stopping or delaying the street cars. The water mains were shut off for only a few hours.

SPECIAL RIVETED RECTANGULAR WATER PIPE, OVER ROOF OF SUBWAY AT 126TH STREET AND LENOX AVENUE

As has been said, the typical subway near the surface was used for about one-half of the road. Since the sewers were at such a depth as to interfere with the construction of the subway, it meant that the sewers along that half had to be reconstructed. This indicates but very partially the magnitude of the sewer work, however, because nearly as many main sewers had to be reconstructed off the route of the subway as on the route; 7.21 miles of main sewers along the route were reconstructed and 5.13 miles of main sewers off the route. The reason why so many main sewers on streets away from the subway had to be rebuilt, was that, from 42d Street, south, there is a natural ridge, and before the construction of the subway sewers drained to the East River and to the North River from the ridge. The route of the subway was so near to the dividing line that the only way to care for the sewers was, in many instances, to build entirely new outfall sewers.

THREE-TRACK CONCRETE ARCH—117TH STREET AND BROADWAY

A notable example of sewer diversion was at Canal Street, where the flow of the sewer was carried into the East River instead of into the Hudson River, permitting the sewer to be bulkheaded on the west side and continued in use. On the east side a new main sewer was constructed to empty into the East River. The new east-side sewer was built off the route of the subway for over a mile. An interesting feature in the construction was the work at Chatham Square, where a 6-1/2-foot circular brick conduit was built. The conjunction at this point of numerous electric surface car lines, elevated railroad pillars, and enormous vehicular street traffic, made it imperative that the surface of the street should not be disturbed, and the sewer was built by tunneling. This tunneling was through very fine running sand and the section to be excavated was small. To meet these conditions a novel method of construction was used. Interlocked poling boards were employed to support the roof and were driven by lever jacks, somewhat as a shield is driven in the shield system of tunneling. The forward ends of the poling boards were supported by a cantilever beam. The sides and front of the excavation were supported by lagging boards laid flat against and over strips of canvas, which were rolled down as the excavation progressed. The sewer was completed and lined in lengths of from 1 foot to 4-1/2 feet, and at the maximum rate of work about 12 feet of sewer were finished per week.

CONSTRUCTION OF FORT GEORGE TUNNEL

At 110th Street and Lenox Avenue a 6-1/2-foot circular brick sewer intersected the line of the subway at a level which necessitated its removal or subdivision. The latter expedient was adopted, and three 42-inch cast-iron pipes were passed under the subway. (See photograph on page 50.) At 149th Street and Railroad Avenue a sewer had to be lowered below tide level in order to cross under the subway. To do this two permanent inverted siphons were built of 48-inch cast-iron pipe. Two were built in order that one might be used, while the other could be shut off for cleaning, and they have proved very satisfactory. This was the only instance where siphons were used. In this connection it is worthy of note that the general changes referred to gave to the city much better sewers as substitutes for the old ones.

A number of interesting methods of providing for subsurface structures are shown in photographs pages 51 to 54. From the General Post-office at Park Row to 28th Street, just below the surface, there is a system of pneumatic mail tubes for postal delivery. Of course, absolutely no change in alignment could be permitted while these tubes were in use carrying mail. It was necessary, therefore, to support them very carefully. The slightest deviation in alignment would have stopped the service.

TWO COLUMN BENT VIADUCT

TRAVELER FOR ERECTING FORMS, CENTRAL PARK
TUNNEL—(IN THIS TUNNEL DUCTS ARE BUILT IN THE
SIDEWALLS)

Between 33d Street and 42d Street under Park Avenue, between 116th Street and 120th Street under Broadway, between 157th Street and Fort George under Broadway and Eleventh Avenue (the second longest double-track rock tunnel in the United States, the Hoosac tunnel being the only one of greater length), and between 104th Street and Broadway under Central Park to Lenox Avenue, the road is in rock tunnel lined with concrete. From 116th Street to 120th Street the tunnel is 37-1/2 feet wide, one of the widest concrete arches in the world. On the section from Broadway and 103d Street to Lenox Avenue and 110th Street under Central Park, a two-track subway was driven through micaceous rock by taking out top headings and then two full-width benches. The work was done from two shafts and one portal. All drilling for the headings was done by an eight-hour night shift, using percussion drills. The blasting was done early in the morning and the day gang removed the spoil, which was hauled to the shafts and the portal in cars drawn by mules. A large part of the rock was crushed for concrete. The concrete floor was the first part of the lining to be put in place. Rails were laid on it for a traveler having moulds attached to its sides, against which the walls were built. A similar traveler followed with the centering for the arch roof, a length of about 50 feet being completed at one operation.

FOUR COLUMN (TOWER) VIADUCT CONSTRUCTION

MANHATTAN VALLEY VIADUCT, LOOKING NORTH

ERECTION OF ARCH, MANHATTAN VALLEY VIADUCT

On the Park Avenue section from 34th Street to 41st Street two separate double-track tunnels were driven below a double-track electric railway tunnel, one on each side. The work was done from four shafts, one at each end of each tunnel. At first, top headings were employed at the north ends of both tunnels and at the south end of the west tunnel; at the south end of the east tunnel a bottom heading was used. Later, a bottom heading was also used at the south end of the west tunnel. The rock was very irregular and treacherous in character, and the strata inclined so as to make the danger of slips a serious one. The two headings of the west tunnel met in February and those of the east tunnel in March, 1902, and the widening of the tunnels to the full section was immediately begun. Despite the adoption of every precaution suggested by experience in such work, some disturbance of the surface above the east tunnel resulted, and several house fronts were damaged. The portion of the tunnel affected was bulkheaded at each end, packed with rubble and grouted with Portland cement mortar injected under pressure through pipes sunk from the street surface above. When the interior was firm, the tunnel was redriven, using much the same methods that are employed for tunnels through earth when the arch lining is built before the central core, or dumpling of earth, is removed. The work had to be done very slowly to prevent any further settlement of the ground, and the completion of the widening of the other parts of the tunnels also proceeded very slowly, because as soon as the slip occurred a large amount of timbering was introduced, which interfered seriously with the operations. After the lining was completed, Portland cement grout was again injected under pressure, through holes left in the roof, until further movement of the fill overhead was absolutely prevented.

COMPLETED ARCH AT MANHATTAN STREET

As has been said, the tunnel between 157th Street and Fort George is the second longest two-track tunnel in the United States. It was built in a remarkably short time, considering the fact that the work was prosecuted from two portal headings and from two shafts. One shaft was at 168th Street and the other at 181st Street, the work proceeding both north and south from each shaft. The method employed for the work (Photograph on page 56) was similar to that used under Central Park. The shafts at 168th Street and at 181st Street were located at those points so that they might be used for the permanent elevator equipment for the stations at these streets. These stations each have an arch span of about 50 feet, lined with brick.

Steel Viaduct

The elevated viaduct construction extends from 125th Street to 133d Street and from Dyckman Street to Bailey Avenue on the western branch, and from Brook and Westchester Avenues to Bronx Park on the eastern, a total distance of about 5 miles. The three-track viaducts are carried on two column bents where the rail is not more than 29 feet above the ground level, and on four-column towers for higher structures. In the latter case, the posts of a tower are 29 feet apart transversely and 20 or 25 feet longitudinally, as a rule, and the towers are from 70 to 90 feet apart on centers. The tops of the towers have X-bracing and the connecting spans have two panels of intermediate vertical sway bracing between the three pairs of longitudinal girders. In the low viaducts, where there are no towers, every fourth panel has zigzag lateral bracing in the two panels between the pairs of longitudinal girders.

PROFILE OF HARLEM RIVER TUNNEL AND APPROACHES

SECTION OF HARLEM RIVER TUNNEL DURING CONSTRUCTION

ASSEMBLING IRON WORK ON PONTOON—HARLEM RIVER TUNNEL

The towers have columns consisting as a rule of a 16 x 7/16-inch web plate and four 6 x 4 x 5/8-inch bulb angles. The horizontal struts in their cross-bracing are made of four 4 x 3-inch angles, latticed to form an I-shaped cross-section. The X-bracing consists of single 5 x 3-1/2-inch angles. The tops of the columns have horizontal cap angles on which are riveted the lower flanges of the transverse girders; the end angles of the girder and the top of the column are also connected by a riveted splice plate. The six longitudinal girders are web-riveted to the transverse girders. The outside longitudinal girder on each side of the viaduct has the same depth across the tower as in the connecting span, but the four intermediate lines are not so deep across the towers. In the single trestle bents the columns are the same as those just described, but the diagonal bracing is replaced by plate knee-braces.

The Manhattan Valley Viaduct on the West Side line, has a total length of 2,174 feet. Its most important feature is a two-hinged arch of 168-1/2 feet span, which carries platforms shaded by canopies, but no station buildings. The station is on the ground between the surface railway tracks. Access to the platforms is obtained by means of escalators. It has three lattice-girder two-hinge ribs 24-1/2 feet apart on centers, the center line of each rib being a parabola. Each half rib supports six spandrel posts carrying the roadway, the posts being seated directly over vertical web members of the rib. The chords of the ribs are 6 feet apart and of an H-section, having four 6 x 6-inch angles and six 15-inch flange and web plates for the center rib and lighter sections for the outside ribs. The arch was erected without false work.

SHOWING CONCRETE OVER IRON WORK—HARLEM RIVER TUNNEL

The viaduct spans of either approach to the arch are 46 to 72 feet long. All transverse girders are 31 feet 4 inches long, and have a 70 x 3/8-inch web plate and four 6 x 4-inch angles. The two outside longitudinal girders of deck spans are 72 inches deep and the other 36 inches. All are 3/8-inch thick and their four flange angles vary in size from 5 x 3-1/2 to 6 x 6 inches, and on the longest spans there are flange plates. At each end of the viaduct there is a through span with 90-inch web longitudinal girders.

Each track was proportioned for a dead load of 330 pounds per lineal foot and a live load of 25,000 pounds per axle. The axle spacing in the truck was 5 feet and the pairs of axles were alternately 27 and 9 feet apart. The traction load was taken at 20 per cent. of the live load, and a wind pressure of 500 pounds per lineal foot was assumed over the whole structure.

Tubes under Harlem River

One of the most interesting sections of the work is that which approaches and passes under the Harlem River, carrying the two tracks of the East Side line. The War Department required a minimum depth of 20 feet in the river at low tide, which fixed the elevation of the roof of the submerged part of the tunnel. This part of the line, 641 feet long, consists of twin single-track cast-iron cylinders 16 feet in diameter enveloped in a large mass of concrete and lined with the same material. The approach on either side is a double-track concrete arched structure. The total length of the section is 1,500 feet.

The methods of construction employed were novel in subaqueous tunneling and are partly shown on photographs on pages 62 and 63. The bed of the Harlem River at the point of tunneling consists of mud, silt, and sand, much of which was so nearly in a fluid condition that it was removed by means of a jet. The maximum depth of excavation was about 50 feet. Instead of employing the usual method of a shield and compressed air at high pressure, a much speedier device was contrived.

The river crossing has been built in two sections. The west section was first built, the War Department having forbidden the closing of more than half the river at one time. A trench was dredged over the line of the tunnel about 50 feet wide and 39 feet below low water. This depth was about 10 feet above the sub-grade of the tunnel. Three rows of piles were next driven on each side of the trench from the west bank to the middle of the river and on them working platforms were built, forming two wharves 38 feet apart in the clear. Piles were then driven over the area to be covered by the subway, 6 feet 4 inches apart laterally and 8 feet longitudinally. They were cut off about 11 feet above the center line of each tube and capped with timbers 12 inches square. A thoroughly-trussed framework was then floated over the piles and sunk on them. The trusses were spaced so as to come between each transverse row of piles and were connected by eight longitudinal sticks or stringers, two at the top and two at the bottom on each side. The four at each side were just far enough apart to allow a special tongue and grooved 12-inch sheet piling to be driven between them. This sheathing was driven to a depth of 10 to 15 feet below the bottom of the finished tunnel.

A well-calked roof of three courses of 12-inch timbers, separated by 2-inch plank, was then floated over the piles and sunk. It had three timber shafts 7 x 17 feet in plan, and when it was in place and covered with earth it formed the top of a caisson with the sheet piling on the sides and ends, the latter being driven after the roof was in place. The excavation below this caisson was made under air pressure, part of the material being blown out by water jets and the remainder removed through the airlocks in the shafts. When the excavation was completed, the piles were temporarily braced and the concrete and cast-iron lining put in place, the piles being cut off as the concrete bed was laid up to them.

The second or eastern section of this crossing was carried on by a modification of the plan just mentioned. Instead of using a temporary timber roof on the side walls, the permanent iron and concrete upper half of the tunnels was employed as a roof for the caisson. The trench was dredged nearly to sub-grade and its sides provided with wharves as before, running out to the completed half of the work. The permanent foundation piles were then driven and a timber frame sunk over them to serve as a guide for the 12-inch sheet piling around the site. Steel pilot piles with water jets were driven in advance of the wood-sheet piles, and if they struck any boulders the latter were drilled and blasted. The steel piles were withdrawn by a six-part tackle and hoisting engine, and then the wooden piles driven in their place.

When the piling was finished, a pontoon 35 feet wide, 106 feet long, and 12 feet deep was built between the wharves, and upon a separate platform or deck on it the upper half of the cast-iron shells were assembled, their ends closed by steel-plate diaphragms and the whole covered with concrete. The pontoon was then submerged several feet, parted at its center, and each half drawn out endwise from beneath the floating top of the tunnel. The latter was then loaded and carefully sunk into place, the connection with the shore section being made by a diver, who entered the roof through a special opening. When it was finally in place, men entered through the shore section and cut away the wood bottom, thus completing the caisson so that work could proceed below it as before. Three of these caissons were required to complete the east end of the crossing.

**LOOKING UP BROADWAY FROM TRINITY CHURCH—
SHOWING WORKING PLATFORM AND GAS MAINS
TEMPORARILY SUPPORTED OVERHEAD**

The construction of the approaches to the tunnel was carried out between heavy sheet piling. The excavation was over 40 feet deep in places and very wet, and the success of the work was largely due to the care taken in driving the 12-inch sheet piling.

Methods of Construction Brooklyn Extension

A number of interesting features should be noted in the methods of construction adopted on the Brooklyn Extension.

The types of construction on the Brooklyn Extension have already been spoken of. They are (1) typical flat-roof steel beam subway from the Post-office, Manhattan, to Bowling Green; (2) reinforced concrete typical subway in Battery Park, Manhattan, and from Clinton Street to the terminus, in Brooklyn; (3) two single track cast-iron-lined tubular tunnels from Battery Park, under the East River, and under Joralemon Street to Clinton Street, Brooklyn.

Under Broadway, Manhattan, the work is through sand, the vehicular and electric street car traffic, the network of subsurface structures, and the high buildings making this one of the most difficult portions of the road to build. The street traffic is so great that it was decided that during the daytime the surface of the street should be maintained in a condition suitable for ordinary traffic. This was accomplished by making openings in the sidewalk near the curb, at two points, and erecting temporary working platforms over the street 16 feet from the surface. The excavations are made by the ordinary drift and tunnel method. The excavated material is hoisted from the openings to the platforms and passed through chutes to wagons. On the street surface, over and in advance of the excavations, temporary plank decks are placed and maintained during the drifting and tunneling operations, and after the permanent subway structure has been erected up to the time when the street surface is permanently restored. The roof of the subway is about 5 feet from the surface of the street, which has made it necessary to care for the gas and water mains. This has been done by carrying the mains on temporary trestle structures over the sidewalks. The mains will be restored to their former position when the subway structure is complete.

From Bowling Green, south along Broadway, State Street and in Battery Park, where the subway is of reinforced concrete construction, the "open cut and cover" method is employed, the elevated and surface railroad structures being temporarily supported by wooden and steel trusses and finally supported by permanent foundations resting on the subway roof. From Battery Place, south along the loop work, the greater portion of the excavation is made below mean high-water level, and necessitates the use of heavy tongue and grooved sheeting and the operation of two centrifugal pumps, day and night.

The tubes under the East River, including the approaches, are each 6,544 feet in length. The tunnel consists of two cast-iron tubes 15-1/2 feet diameter inside, the lining being constructed of cast-iron plates, circular in shape, bolted together and reinforced by grouting outside of the plates and beton filling on the inside to the depth of the flanges. The tubes are being constructed under air pressure through solid rock from the Manhattan side to the middle of the East River by the ordinary rock tunnel drift method, and on the Brooklyn side through sand and silt by the use of hydraulic shields. Four shields have been installed, weighing 51 tons each. They are driven by hydraulic pressure of about 2,000 tons. The two shields drifting to the center of the river from Garden Place are in water-bearing sand and are operated under air pressure. The river tubes are on a 3.1 per cent. grade and in the center of the river will reach the deepest point, about 94 feet below mean high-water level.

The typical subway of reinforced concrete from Clinton Street to the Flatbush Avenue terminus is being constructed by the method commonly used on the Manhattan-Bronx route. From Borough Hall to the terminus the route of the subway is directly below an elevated railway structure, which is temporarily supported by timber bracing, having its bearing on the street surface and the tunnel timbers. The permanent support will be masonry piers built upon the roof of the subway structure. Along this portion of the route are street surface electric roads, but they are operated by overhead trolley and the tracks are laid on ordinary ties. It has, therefore, been much less difficult to care for them during the construction of the subway. Work is being prosecuted on the Brooklyn Extension day and night, and in Brooklyn the excavation is made much more rapidly by employing the street surface trolley roads to remove the excavated material. Spur tracks have been built and flat cars are used, much of the removal being done at night.

CHAPTER III

POWER HOUSE BUILDING

The power house is situated adjacent to the North River on the block bounded by West 58th Street, West 59th Street, Eleventh Avenue, and Twelfth Avenue. The plans were adopted after a thorough study by the engineers of Interborough Rapid Transit Company of all the large power houses already completed and of the designs of the large power houses in process of construction in America and abroad. The building is large, and when fully equipped it will be capable of producing more power than any electrical plant ever built, and the study of the designs of other power houses throughout the world was pursued with the principal object of reducing to a minimum the possibility of interruption of service in a plant producing the great power required.

The type of power house adopted provides for a single row of large engines and electric generators, contained within an operating room placed beside a boiler house, with a capacity of producing, approximately, not less than 100,000 horse power when the machinery is being operated at normal rating.

Location and General Plan of Power House

The work of preparing the detailed plans of the power house structure was, in the main, completed early in 1902, and resulted in the present plan, which may briefly be described as follows: The structure is divided into two main parts—an operating room and a boiler house, with a partition wall between the two sections. The face of the structure on Eleventh Avenue is 200 feet wide, of which width the boiler house takes 83 feet and the operating section 117 feet. The operating room occupies the northerly side of the structure and the boiler house the southerly side. The designers were enabled to employ a contour of roof and wall section for the northerly side that was identical with the roof and wall contour of the southerly side, so that the building, when viewed from either end, presents a symmetrical appearance with both sides of the building alike in form and design. The operating room section is practically symmetrical in its structure, with respect to its center; it consists of a central area, with a truss roof over same along with galleries at both sides. The galleries along the northerly side are primarily for the electrical apparatus, while those along the southerly side are given up chiefly to the steam-pipe equipment. The boiler room section is also practically symmetrical with respect to its center.

A sectional scheme of the power house arrangement was determined on, by which the structure was to consist of five generating sections, each similar to the others in all its mechanical details; but, at a later date, a sixth section was added, with space on the lot for a seventh section. Each section embraces one chimney along with the following generating equipment:— twelve boilers, two engines, each direct connected to a 5,000 kilowatt alternator; two condensing equipments, two boiler-feed pumps, two smoke-flue systems, and detail apparatus necessary to make each section complete in itself. The only variation is the turbine plant hereafter referred to. In addition to the space occupied by the sections, an area was set aside, at the Eleventh Avenue end of the structure, for the passage of the railway spur from the New York Central tracks. The total length of the original five-section power house was 585 feet 9-1/2 inches, but the additional section afterwards added makes the over all length of the structure 693 feet 9-3/4 inches. In the fourth section it was decided to omit a regular engine with its 5,000 kilowatt generator, and in its place substitute a 5,000 kilowatt lighting and exciter outfit. Arrangements were made, however, so that this outfit can afterward be replaced by a regular 5,000 kilowatt traction generator.

CROSS SECTION OF POWER HOUSE IN PERSPECTIVE

The plan of the power station included a method of supporting the chimneys on steel columns, instead of erecting them through the building, which modification allowed for the disposal of boilers in spaces which would otherwise be occupied by the chimney bases. By this arrangement it was possible to place all the boilers on one floor level. The economizers were placed above the boilers, instead of behind them, which made a material saving in the width of the boiler room. This saving permitted the setting aside of the aforementioned gallery at the side of the operating room, closed off from both boiler and engine rooms, for the reception of the main-pipe systems and for a pumping equipment below it.

The advantages of the plan can be enumerated briefly as follows: The main engines, combined with their alternators, lie in a single row along the center line of the operating room with the steam or operating end of each engine facing the boiler house and the opposite end toward the electrical switching and controlling apparatus arranged along the outside wall. Within the area between the boiler house and operating room there is placed, for each engine, its respective complement of pumping apparatus, all controlled by and under the operating jurisdiction of the engineer for that engine. Each engineer has thus full control of the pumping machinery required for his unit. Symmetrically arranged with respect to the center line of each engine are the six boilers in the boiler room, and the piping from these six boilers forms a short connection between the nozzles on the boilers and the throttles on the engine. The arrangement of piping is alike for each engine, which results in a piping system of maximum simplicity that can be controlled, in the event of difficulty, with a degree of certainty not possible with a more complicated system. The main parts of the steam-pipe system can be controlled from outside this area.

The single tier of boilers makes it possible to secure a high and well ventilated boiler room with ventilation into a story constructed above it, aside from that afforded by the windows themselves. The boiler room will therefore be cool in warm weather and light, and all difficulties from escaping steam will be minimized. In this respect the boiler room will be superior to corresponding rooms in plants of older construction, where they are low, dark, and often very hot during the summer season. The placing of the economizers, with their auxiliary smoke flue connections, in the economizer room, all symmetrically arranged with respect to each chimney, removes from the boiler room an element of disturbance and makes it possible to pass directly from the boiler house to the operating room at convenient points along the length of the power house structure. The location of each chimney in the center of the boiler house between sets of six boilers divides the coal bunker construction into separate pockets by which trouble from spontaneous combustion can be localized, and, as described later, the divided coal bunkers can provide for the storage of different grades of coal. The unit basis on which the economizer and flue system is constructed will allow making repairs to any one section without shutting off the portions not connected directly to the section needing repair.

The floor of the power house between the column bases is a continuous mass of concrete nowhere less than two feet thick. The massive concrete foundations for the reciprocating engines contain each 1,400 yards of concrete above mean high water level, and in some cases have twice as much below that point. The total amount of concrete in the foundations of the finished power house is about 80,000 yards.

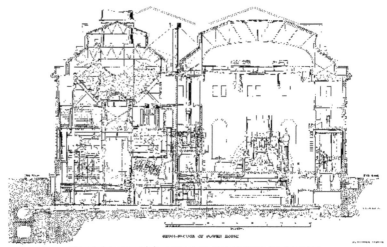

CROSS-SECTION OF POWER HOUSE

Water for condensing purposes is drawn from the river and discharged into it through two monolithic concrete tunnels parallel to the axis of the building. The intake conduit has an oval interior, 10 x 8-1/2 feet in size, and a rectangular exterior cross-section; the outflow tunnel has a horseshoe-shape cross-section and is built on top of the intake tunnel. These tunnels were built throughout in open trench, which, at the shore end, was excavated in solid rock. At the river end the excavation was, at some places, almost entirely through the fill and mud and was made in a cofferdam composed chiefly of sheet piles. As it was impossible to drive these piles across the old timber crib which formed the old dock front, the latter was cut through by a pneumatic caisson of wooden-stave construction, which formed part of one side of the cofferdam. At the river end of the cofferdam the rock was so deep that the concrete could not be carried down to its surface, and the tunnel section was built on a foundation of piles driven to the rock and cut off by a steam saw 19-1/2 feet below mean hightide. This section of the tunnel was built in a 65 x 48-foot floating caisson 24 feet deep. The concrete was rammed in it around the moulds and the sides were braced as it sunk. After the tunnel sections were completed, the caisson was sunk, by water ballast, to a bearing on the pile foundation.

Adjacent to the condensing water conduits is the 10 x 15-foot rectangular concrete tunnel, through which the underground coal conveyor is installed between the shore end of the pier and the power house.

Steel Work

The steel structure of the power house is independent of the walls, the latter being self-supporting and used as bearing walls only for a few of the beams in the first floor. Although structurally a single building, in arrangement it is essentially two, lying side by side and separated by a brick division wall.

There are 58 transverse and 9 longitudinal rows of main columns, the longitudinal spacing being 18 feet and 36 feet for different rows, with special bracing in the boiler house to accommodate the arrangement of boilers. The columns are mainly of box section, made up of rolled or built channels and cover plates. They are supported by cast-iron bases, resting on the granite capstones of the concrete foundation piers.

Both the boiler house and the engine house have five tiers of floor framing below the flat portion of the roof, the three upper tiers of the engine house forming galleries on each side of the operating room, which is clear for the full height of the building.

The boiler house floors are, in general, framed with transverse plate girders and longitudinal rolled beams, arranged to suit the particular requirements of the imposed loads of the boilers, economizers, coal, etc., while the engine-room floors and pipe and switchboard galleries are in general framed with longitudinal plate girders and transverse beams.

There are seven coal bunkers in the boiler house, of which five are 77 feet and two 41 feet in length by 60 feet in width at the top, the combined maximum capacity being 18,000 tons. The bunkers are separated from each other by the six chimneys spaced along the center line of the boiler house. The bottom of the bunkers are at the fifth floor, at an elevation of about 66 feet above the basement. The bunkers are constructed with double, transverse, plate girder frames at each line of columns, combined with struts and ties, which balance the outward thrust of the coal against the sides. The frames form the outline of the bunkers with slides sloping at 45 degrees, and carry longitudinal I-beams, between which are built concrete arches, reinforced with expanded metal, the whole surface being filled with concrete over the tops of the beams and given a two-inch granolithic finish.

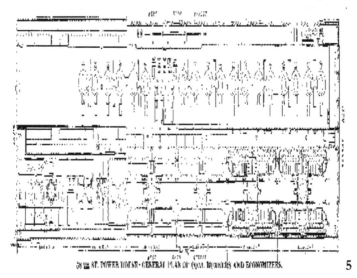

8TH ST. POWER HOUSE—GENERAL PLAN OF COAL BUNKERS AND ECONOMIZERS.

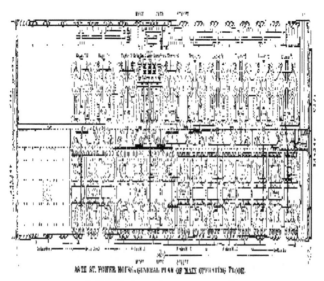

58TH ST. POWER HOUSE—GENERAL PLAN OF MAIN OPERATING FLOOR.

The six chimneys, spaced 108 feet apart, and occupying the space between the ends of the adjacent coal bunkers, are supported on plate-girder platforms in the fifth floor, leaving the space below clear for a symmetrical arrangement of the boilers and economizers from end to end of the building. The platforms are framed of single-web girders 8 feet deep, thoroughly braced and carrying on their top flanges a grillage of 20-inch I-beam. A system of bracing for both the chimney platforms and coal bunkers is carried down to the foundations in traverse planes about 30 feet apart.

The sixth tier of beams constitute a flat roof over a portion of the building at the center and sides. In the engine room, at this level, which is 64 feet above the engine-room floor, are provided the two longitudinal lines of crane runway girders upon which are operated the engine-room cranes. Runways for 10-ton hand cranes are also provided for the full length of the boiler room, and for nearly the full length of the north panel in the engine room.

Some of the loads carried by the steel structure are as follows: In the engine house, operating on the longitudinal runways as mentioned, are one 60-ton and one 25-ton electric traveling crane of 75 feet span. The imposed loads of the steam-pipe galleries on the south side and the switchboard galleries on the north side are somewhat irregularly distributed, but are equivalent to uniform loads of 250 to 400 pounds per square foot. In the boiler house the weight of coal carried is about 45 tons per longitudinal foot of the building; the weight of the brick chimneys is 1,200 tons each; economizers, with brick setting, about 4-1/2 tons per longitudinal foot; suspended weight of the boilers 96 tons each, and the weight of the boiler setting, carried on the first floor framing, 160 tons each. The weight of structural steel used in the completed building is about 11,000 tons.

Power House Superstructure

The design of the facework of the power house received the personal attention of the directors of the company, and its character and the class of materials to be employed were carefully considered. The influence of the design on the future value of the property and the condition of the environment in general were studied, together with the factors relating to the future ownership of the plant by the city. Several plans were taken up looking to the construction of a power house of massive and simple design, but it was finally decided to adopt an ornate style of treatment by which the structure would be rendered architecturally attractive and in harmony with the recent tendencies of municipal and city improvements from an architectural standpoint. At the initial stage of the power house design Mr.

Stanford White, of the firm of McKim, Mead & White, of New York, volunteered his services to the company as an adviser on the matter of the design of the facework, and, as his offer was accepted, his connection with the work has resulted in the development of the present exterior design and the selection of the materials used.

The Eleventh Avenue façade is the most elaborately treated, but the scheme of the main façade is carried along both the 58th and 59th Street fronts. The westerly end of the structure, facing the river, may ultimately be removed in case the power house is extended to the Twelfth Avenue building line for the reception of fourteen generating equipments; and for this reason this wall is designed plainly of less costly material.

The general style of the facework is what may be called French Renaissance, and the color scheme has, therefore, been made rather light in character. The base of the exterior walls has been finished with cut granite up to the water table, above which they have been laid up with a light colored buff pressed brick. This brick has been enriched by the use of similarly colored terra-cotta, which appears in the pilasters, about the windows, in the several entablatures, and in the cornice and parapet work. The Eleventh Avenue façade is further enriched by marble medallions, framed with terra-cotta, and by a title panel directly over the front of the structure.

The main entrance to the structure is situated at its northeast corner, and, as the railroad track passes along just inside the building, the entrance proper is the doorway immediately beyond the track, and opens into the entrance lobby. The doorway is trimmed with cut granite and the lobby is finished with a marble wainscoting.

The interior of the operating room is faced with a light, cream-colored pressed brick with an enameled brick wainscoting, eight feet high, extending around the entire operating area; the wainscoting is white except for a brown border and base. The offices, the toilets and locker rooms are finished and fitted with materials in harmony with the high-class character of the building. The masonry-floor construction consists of concrete reinforced with expanded metal, and except where iron or other floor plates are used, or where tile or special flooring is laid, the floor is covered with a hard cement granolithic finish.

In the design of the interior arrangements, the value of a generous supply of stairways was appreciated, in order that all parts of the structure might be made readily accessible, especially in the boiler house section. In the boiler house and machinery portion of the plant the stairways, railings, and accessories are plainly but strongly constructed. The main stairways are, however, of somewhat ornate design, with marble and other trim work, and the railings of the main gallery construction are likewise of ornate treatment. All exterior doors and trim are of metal and all interior carpenter work is done with Kalomein iron protection, so that the building, in its strictest sense, will contain no combustible material.

Chimneys

The complete 12-unit power house will have six chimneys, spaced 108 feet apart on the longitudinal center line of the boiler room, each chimney being 15 feet in inside diameter at the top, which is 225 feet above the grate bars. Each will serve the twelve boilers included in the section of which it is the center, these boilers having an aggregate of 72,000 square feet of heating surface. By these dimensions each chimney has a fair surplus capacity, and it is calculated that, with economizers in the path of the furnace gases, there will be sufficient draft to meet a demand slightly above the normal rating of the boilers. To provide for overload capacity, as may be demanded by future conditions, a forced draft system will be supplied, as described later.

As previously stated, the chimneys are all supported upon the steel structure of the building at an elevation of 76 feet above the basement floor and 63 feet above the grates. The supporting platforms are, in each case, carried on six of the building columns (the three front columns of two groups of boilers on opposite sides of the center aisle of the boiler room), and each platform is composed of single-web plate girders, well braced and surmounted by a grillage of 20-inch I-beams. The grillage is filled solidly with concrete and flushed smooth on top to receive the brickwork of the chimney.

Each chimney is 162 feet in total height of brickwork above the top of the supporting platform, and each chimney is 23 feet square in the outside dimension at the base, changing to an octagonal form at a point 14 feet 3 inches above the base. This octagonal form is carried to a height of 32 feet 6 inches above the base, at which point the circular section of radial brick begins.

The octagonal base of the chimney is of hard-burned red brick three feet in thickness between the side of the octagon and the interior circular section. The brick work is started from the top of the grillage platform with a steel channel curb, three feet in depth, through which two lines of steel rods are run in each direction, thus binding together the first three feet of brickwork, and designed to prevent any flaking at the outside. At a level of three feet above the bottom of the brickwork, a layer of water-proofing is placed over the interior area and covered with two courses of brick, upon which are built diagonal brick walls, 4 inches thick, 12 inches apart, and about 18 inches in height. These walls are themselves perforated at intervals, and the whole is covered with hand-burned terra-cotta blocks, thus forming a cellular air space, which communicates with the exterior air and serves as an insulation against heat for the steelwork beneath. A single layer of firebrick completes the flooring of the interior area, which is also flush with the bottom of the flue openings.

There are two flue openings, diametrically opposite, and 6 feet wide by 17 feet high to the crown of the arched top. They are lined with fire brick, which joins the fire-brick lining of the interior of the shaft, this latter being bonded to the red-brick walls to a point 6 feet below the top of the octagon, and extended above for a height of 14 feet within the circular shaft, as an inner shell. The usual baffle wall is provided of fire brick, 13 inches thick, extending diagonally across the chimney, and 4 feet above the tops of the flue openings.

Where the chimney passes through the roof of the boiler house, a steel plate and angle curb, which clears the chimney by 6 inches at all points, is provided in connection with the roof framing. This is covered by a hood flashed into the brickwork, so that the roof has no connection with or bearing upon the chimney.

At a point 4 feet 6 inches below the cap of the chimney the brickwork is corbeled out for several courses, forming a ledge, around the outside of which is placed a wrought-iron railing, thus forming a walkway around the circumference of the chimney top. The cap is of cast iron, surmounted by eight 3 x 1-inch wrought-iron ribs, bent over the outlet and with pointed ends gathered together at the center. The lightning conductors are carried down the outside of the shaft to the roof and thence to the ground outside of the building. Galvanized iron ladder rungs were built in the brickwork, for ladders both inside and outside the shaft.

The chimneys, except for the octagonal red-brick base, are constructed of the radial perforated bricks. The lightning rods are tipped with pointed platinum points about 18 inches long.

North River Pier

Exceptional facilities have been provided for the unloading of coal from vessels, or barges, which can be brought to the northerly side of the recently constructed pier at the foot of West 58th Street. The pier was specially built by the Department of Docks and Ferries and is 700 feet long and 60 feet wide.

The pier construction includes a special river wall across 58th Street at the bulkhead line through which the condensing water will be taken from and returned to the river. Immediately outside the river wall and beneath the deck of the pier, there is a system of screens through which the intake water is passed. On each side where the water enters the screen chamber, is a heavy steel grillage; inside this is a system of fine screens arranged so that the several screens can be raised, by a special machine, for the purpose of cleaning. The advantages of a well-designed screening outfit has been appreciated, and considerable care has been exercised to make it as reliable and effective as possible.

At each side of the center of the pier, just below the deck, there are two discharge water conduits constructed of heavy timber, to conduct the warm water from the condensers away from the cold water intakes at the screens. Two water conduits are employed, in order that one may be repaired or renewed while using the other; in fact, the entire pier is constructed with the view of renewal without interference in the operation for which it was provided.

CHAPTER IV

POWER PLANT FROM COAL PILE TO SHAFTS OF ENGINES AND TURBINES

From the minute and specific description in Chapter III, a clear idea will have been obtained of the power house building and its adjuncts, as well as of the features which not only go to make it an architectural landmark, but which adapt it specifically for the vital function that it is called upon to perform. We now come to a review and detailed description of the power plant equipment in its general relation to the building, and "follow the power through" from the coal pile to the shafts of the engines or steam turbines attached to the dynamos which generate current for power and for light.

Coal and Ash Handling Equipment

The elements of the coal handling equipment comprise a movable electric hoisting tower with crushing and weighing apparatus—a system of horizontal belt conveyors, with 30-inch belts, to carry the crushed and weighed coal along the dock and thence by tunnel underground to the southwest corner of the power house; a system of 30-inch belt conveyors to elevate the coal a distance of 110 feet to the top of the boiler house, at the rate of 250 tons per hour or more, if so desired, and a system of 20-inch belt conveyors to distribute it horizontally over the coal bunkers. These conveyors have automatic self reversing trippers, which distribute the coal evenly in the bunkers. For handling different grades of coal, distributing conveyors are arranged underneath the bunkers for delivering the coal from a particular bunker through gates to the downtake hoppers in front of the boilers, as hereafter described.

The equipment for removing ashes from the boiler room basement and for storing and delivering the ashes to barges, comprises the following elements: A system of tracks, 24 inches gauge, extending under the ash-hopper gates in the boiler-house cellar and extending to an elevated storage bunker at the water front. The rolling stock consists of 24 steel cars of 2 tons capacity, having gable bottoms and side dumping doors. Each car has two four-wheel pivoted trucks with springs. Motive power is supplied by an electric storage battery locomotive. The cars deliver the ashes to an elevating belt conveyor, which fills the ash bunker. This will contain 1,000 tons, and is built of steel with a suspension bottom lined with concrete. For delivering stored ashes to barges, a collecting belt extends longitudinally under the pocket, being fed by eight gates. It delivers ashes to a loading belt conveyor, the outboard end of which is hinged so as to vary the height of delivery and to fold up inside the wharf line when not in use.

The coal handling system in question was adopted because any serious interruption of service would be of short duration, as any belt, or part of the belt mechanism, could quickly be repaired or replaced. The system also possessed advantages with respect to the automatic even distribution of coal in the bunkers, by means of the self reversing trippers. These derive their power from the conveying belts. Each conveyor has a rotary cleaning brush to cleanse the belt before it reaches the driving pulley and they are all driven by induction motors.

The tower frame and boom are steel. The tower rolls on two rails along the dock and is self-propelling. The lift is unusually short; for the reason that the weighing apparatus is removed horizontally to one side in a separate house, instead of lying vertically below the crusher. This arrangement reduces by 40 per cent. the lift of the bucket, which is of the clam-shell type of forty-four cubic feet capacity. The motive power for operating the bucket is perhaps the most massive and powerful ever installed for such service. The main hoist is directly connected to a 200 horse-power motor with a special system of control. The trolley engine for hauling the bucket along the boom is also direct coupled to a multipolar motor.

The receiving hopper has a large throat, and a steel grizzly in it which sorts out coal small enough for the stokers and bypasses it around the crusher. The crusher is of the two-roll type, with relieving springs, and is operated by a motor, which is also used for propelling the tower. The coal is weighed in duplex two-ton hoppers.

Special attention has been given to providing for the comfort and safety of the operators. The cabs have baywindow fronts, to enable the men to have an unobstructed view of the bucket at all times without peering through slots in the floor. Walks and hand lines are provided on both sides of the boom for safe inspection. The running ropes pass through hardwood slides, which cover the slots in the engine house roof to exclude rain and snow.

This type of motive power was selected in preference to trolley locomotives for moving the ash cars, owing to the rapid destruction of overhead lines and rail bonds by the action of ashes and water. The locomotive consists of two units, each of which has four driving wheels, and carries its own motor and battery. The use of two units allows the locomotive to round curves with very small overhangs, as compared with a single-body locomotive. Curves of 12 feet radius can be turned with ease. The gross weight of the locomotive is about five tons, all of which is available for traction.

Coal Downtakes

The coal from the coal bunkers is allowed to flow down into the boiler room through two rows of downtakes, one on each side of the central gangway or firing place. Each bunker has eight cast-iron outlets, four on each side, and to these outlets are bolted gate valves for shutting off the coal from the corresponding downtakes. From these gates the downtakes lead to hoppers which are on the economizer floor, and from these hoppers the lower sets of downtakes extend down to the boilers.

Just above the hoppers on the economizer floor the coal downtakes are provided with valves and chutes to feed the coal, either into the hopper or into the distributing flight conveyor alongside of it. These distributing conveyors, one corresponding with each row of downtakes, permits the feeding of coal from any bunker or bunkers to all the boilers when desired. They are the ordinary type of flight conveyor, capable of running in either direction and provided with gates in the bottom of the trough for feeding into the several above mentioned hoppers. In order to eliminate the stresses that would develop in a conveyor of the full length of the building, the conveyors are of half the entire length, with electric driving engines in the center of each continuous line. The installation of this conveyor system, in connection with the coal downtakes, makes it possible to carry a high-grade coal in some of the bunkers for use during periods of heavy load and a cheaper grade in other bunkers for the periods of light load.

To provide means for shutting off the coal supply to each boiler, a small hopper is placed just over each boiler, and the downtake feeding into it is provided with a gate at its lower end. Two vertical downtakes extend down from the boiler hopper to the boiler room floor or to the stokers, as the case may be, and they are hinged just below the boiler hopper to allow their being drawn up out of the way when necessary to inspect the boiler tubes.

WEST END POWER HOUSE IN COURSE OF ERECTION

Wherever the direction of flow of the coal is changed, poke holes are provided in the downtakes to enable the firemen to break any arching tendency of the coal in the downtakes. All parts of the downtakes are of cast iron, except the vertical parts in front of the boilers, which are of wrought-iron pipe. These vertical downtakes are 10 inches in inside diameter, while all others are 14 inches in inside diameter.

Main Boiler Room

The main boiler room is designed to receive ultimately seventy-two safety water tube three drum boilers, each having 6,008 square feet of effective heating surface, by which the aggregate heating surface of the boiler room will be 432,576 square feet.

There are fifty-two boilers erected in pairs, or batteries, and between each battery is a passageway five feet wide. The boilers are designed for a working steam pressure of 225 pounds per square inch and for a hydraulic test pressure of 300 pounds per square inch. Each boiler is provided with twenty-one vertical water tube sections, and each section is fourteen tubes high. The tubes are of lap welded, charcoal iron, 4 inches in diameter and 18 feet long. The drums are 42 inches in diameter and 23 feet and 10 inches long. All parts are of open-hearth steel; the shell plates are 9/16 of an inch thick and the drum head plates 11/16 inch, and in this respect the thickness of material employed is slightly in excess of standard practice. Another advance on standard practice is in the riveting of the circular seams, these being lap-jointed and double riveted. All longitudinal seams are butt-strapped, inside and outside, and secured by six rows of rivets. Manholes are only provided for the front heads, and each front head is provided with a special heavy bronze pad, for making connection to the stop and check feed water valve.

OPERATING ROOM SHOWING CONDENSERS—POWER HOUSE

The setting of the boiler embodies several special features which are new in boiler erection. The boilers are set higher up from the floor than in standard practice, the center of the drums being 19 feet above the floor line. This feature provides a higher combustion chamber, for either hand-fired grates or automatic stokers; and for inclined grate stokers the fire is carried well up above the supporting girders under the side walls, so that these girders will not be heated by proximity to the fire.

As regards the masonry setting, practically the entire inside surface exposed to the hot gases is lined with a high grade of fire brick. The back of the setting, where the rear cleaning is done, is provided with a sliding floor plate, which is used when the upper tubes are being cleaned. There is also a door at the floor line and another at a higher level for light and ventilation when cleaning. Over the tubes arrangements have been made for the reception of superheating apparatus without changing the brickwork. Where the brick walls are constructed, at each side of the building columns at the front, cast-iron plates are erected to a height of 8 feet on each side of the column. An air space is provided between each cast-iron plate and the column, which is accessible for cleaning from the boiler front; the object of the plates and air space being to prevent the transmission of heat to the steel columns.

An additional feature of the boiler setting consists in the employment of a soot hopper, back of each bridge wall, by which the soot can be discharged into ash cars in the basement. The main ash hoppers are constructed of 1/2-inch steel plate, the design being a double inverted pyramid with an ash gate at each inverted apex. The hoppers are well provided with stiffening angles and tees, and the capacity of each is about 80 cubic feet.

In front of all the boilers is a continuous platform of open-work cast-iron plates, laid on steel beams, the level of the platform being 8 feet above the main floor. The platform connects across the firing area, opposite the walk between the batteries, and at these points this platform is carried between the boiler settings. At the rear of the northerly row of boilers the platform runs along the partition wall, between the boiler house and operating room and at intervals doorways are provided which open into the pump area. The level of the platform is even with that of the main operating room floor, so that it may be freely used by the water tenders and by the operating engineers without being obstructed by the firemen or their tools. The platform in front of the boilers will also be used for cleaning purposes, and, in this respect, it will do away with the unsightly and objectionable scaffolds usually employed for this work. The water tenders will also be brought nearer to the water columns than when operating on the main floor. The feed-water valves will be regulated from the platform, as well as the speed of the boiler-feed pumps.

Following European practice, each boiler is provided with two water columns, one on each outside drum, and each boiler will have one steam gauge above the platform for the water tenders and one below the platform for the firemen. The stop and check valves on each boiler drum have been made specially heavy for the requirements of this power house, and this special increase of weight has been applied to all the several minor boiler fittings.

Hand-fired grates of the shaking pattern have been furnished for thirty-six boilers, and for each of these grates a special lower front has been constructed. These fronts are of sheet steel, and the coal passes down to the floor through two steel buckstays which have been enlarged for the purpose. There are three firing doors and the sill of each door is 36 inches above the floor. The gate area of the hand-fired grates is 100 square feet, being 8 feet deep by 12 feet 6 inches wide.

The twelve boilers, which will receive coal from the coal bunker located between the fourth and fifth chimneys, have been furnished with automatic stokers.

It is proposed to employ superheaters to the entire boiler plant.

The boiler-room ceiling has been made especially high, and in this respect the room differs from most power houses of similar construction. The distance from the floor to the ceiling is 35 feet, and from the floor plates over the boilers to the ceiling is 13 feet. Over each boiler is an opening to the economizer floor above, covered with an iron grating. The height of the room, as well as the feature of these openings, and the stairway wells and with the large extent of window opening in the south wall, will make the room light and especially well ventilated. Under these conditions the intense heat usually encountered over boilers will largely be obviated.

In addition to making provisions for the air to escape from the upper part of the boiler room, arrangements have been provided for allowing the air to enter at the bottom. This inflow of air will take place through the southerly row of basement windows, which extend above the boiler room floor, and through the wrought-iron open-work floor construction extending along in the rear of the northerly row of boilers.

A noteworthy feature of the boiler room is the 10-ton hand-power crane, which travels along in the central aisle through the entire length of the structure. This crane is used for erection and for heavy repair, and its use has greatly assisted the speedy assembling of the boiler plant.

Blowers and Air Ducts

In order to burn the finer grades of anthracite coal in sufficient quantities to obtain boiler rating with the hand-fired grates, and in order to secure a large excess over boiler rating with other coals, a system of blowers and air ducts has been provided in the basement under the boilers. One blower is selected for every three boilers, with arrangements for supplying all six boilers from one blower.

The blowers are 11 feet high above the floor and 5 feet 6 inches wide at the floor line. Each blower is direct-connected to a two crank 7-1/2 x 13 x

6-1/2-inch upright, automatic, compound, steam engine of the self-enclosed type, and is to provide a sufficient amount of air to burn 10,000 pounds of combustible per hour with 2 inches of water pressure in the ash pits.

Smoke Flues and Economizers

The smoke flue and economizer construction throughout the building is of uniform design, or, in other words, the smoke flue and economizer system for one chimney is identical with that for every other chimney. In each case, the system is symmetrically arranged about its respective chimney, as can be seen by reference to the plans.

The twelve boilers for each chimney are each provided with two round smoke uptakes, which carry the products of combustion upward to the main smoke flue system on the economizer floor. A main smoke flue is provided for each group of three boilers, and each pair of main smoke flues join together on the center line of the chimney, where in each case one common flue carries the gases into the side of the chimney. The two common flues last mentioned enter at opposite sides of the chimney. The main flues are arranged and fitted with dampers, so that the gases can pass directly to the chimney, or else they can be diverted through the economizers and thence reach the chimney.

The uptakes from each boiler are constructed of 3/8-inch plate and each is lined with radial hollow brick 4 inches thick. Each is provided with a damper which operates on a shaft turning in roller bearings. The uptakes rest on iron beams at the bottom, and at the top, where they join the main flue, means are provided to take up expansion and contraction.

The main flue, which rests on the economizer floor, is what might be called a steel box, constructed of 3/8-inch plate, 6 feet 4 inches wide and 13 feet high. The bottom is lined with brick laid flat and the sides with brick walls 8 inches thick, and the top is formed of brick arches sprung between.

Steam Piping

The sectional plan adopted for the power house has made a uniform and simple arrangement of steam piping possible, with the piping for each section, except that of the turbine bay, identical with that for every other section. Starting with the six boilers for one main engine, the steam piping may be described as follows: A cross-over pipe is erected on each boiler, by means of which and a combination of valves and fittings the steam may be passed through the superheater. In the delivery from each boiler there is a quick-closing 9-inch valve, which can be closed from the boiler room floor by hand or from a distant point individually or in groups of six. Risers with 9-inch wrought-iron goose necks connect each boiler to the steam main, where 9-inch angle valves are inserted in each boiler connection. These valves can be closed from the platform over the boilers, and are grouped three over one set of three boilers and three over the opposite set.

The main from the six boilers is carried directly across the boiler house in a straight line to a point in the pipe area where it rises to connect to the two 14-inch steam downtakes to the engine throttles. At this point the steam can also be led downward to a manifold to which the compensating tie lines are connected. These compensating lines are run lengthwise through the power house for the purpose of joining the systems together, as desired. The two downtakes to the engine throttles drop to the basement, where each, through a goose neck, delivers into a receiver and separating tank and from the tank through a second goose neck into the corresponding throttle.

A quick-closing valve appears at the point where the 17-inch pipe divides into the two 14-inch downtakes and a similar valve is provided at the point where the main connects to the manifold. The first valve will close the steam to the engine and the second will control the flow of steam to and from the manifold. These valves can be operated by hand from a platform located on the wall inside the engine room, or they can be closed from a distant point by hydraulic apparatus. In the event of accident the piping to any engine can be quickly cut out or that system of piping can quickly be disconnected from the compensating system.

The pipe area containing, as mentioned, the various valves described, together with the manifolds and compensating pipes, is divided by means of cross-walls into sections corresponding to each pair of main engines. Each section is thus separated from those adjoining, so that any escape of steam in one section can be localized and, by means of the quick-closing valves, the piping for the corresponding pair of main engines can be disconnected from the rest of the power house.

**VIEW FROM TOP OF CHIMNEY SHOWING WATER
FRONTAGE—POWER HOUSE**

All cast iron used in the fittings is called air-furnace iron, which is a semi-steel and tougher than ordinary iron. All line and bent pipe is of wrought iron, and the flanges are loose and made of wrought steel. The shell of the pipe is bent over the face of the flange. All the joints in the main steam line, above 2-1/2 inches in size, are ground joints, metal to metal, no gaskets being used.

Unlike the flanges ordinarily used in this country, special extra strong proportions have been adopted, and it may be said that all flanges and bolts used are 50 per cent. heavier than the so-called extra heavy proportions used in this country.

Water Piping

The feed water will enter the building at three points, the largest water service being 12 inches in diameter, which enters the structure at its southeast corner. The water first passes through fish traps and thence through meters, and from them to the main reservoir tanks, arranged along the center of the boiler house basement. The water is allowed to flow into each tank by means of an automatic float valve. The water will be partly heated in these reservoir tanks by means of hot water discharged from high-pressure steam traps. In this way the heat contained in the drainage from the high-pressure steam is, for the most part, returned to the boilers. From the reservoir tanks the water is conducted to the feed-water pumps, by which it is discharged through feed-water heaters where it is further heated by the exhaust steam from the condensing and feed-water pumps. From the feed-water heaters the water will be carried direct to the boilers; or through the economizer system to be further heated by the waste gases from the boilers.

PORTION OF MAIN STEAM PIPING IN PIPE AREA

Like the steam-pipe system, the feed-water piping is laid out on the sectional plan, the piping for the several sections being identical, except for the connections from the street service to the reservoir tanks. The feed-water piping is constructed wholly of cast iron, except the piping above the floor line to the boilers, which is of extra heavy semi-annealed brass with extra heavy cast-iron fittings.

Engine and Turbine Equipment

The engine and turbine equipment under contract embraces nine 8,000 to 11,000 horse power main engines, direct-connected to 5,000 kilowatt generators, three steam turbines, direct-connected to 1,875 kilowatt lighting generators and two 400 horse power engines, direct-connected to 250 kilowatt exciter generators.

Main Engines

The main engines are similar in type to those installed in the 74th Street power house of the Manhattan Division of the Interborough Rapid Transit Company, i. e., each consists of two component compound engines, both connected to a common shaft, with the generator placed between the two component engines. The type of engine is now well known and will not be described in detail, but as a comparison of various dimensions and features of the Manhattan and Rapid Transit engines may be of interest, the accompanying tabulation is submitted:

	Manhattan.	Rapid Transit.
Diameter of high-pressure cylinders, inches,	44	42
Diameter of low-pressure cylinders, inches,	88	86
Stroke, inches,	60	60
Speed, revolutions per minute,	75	75
Steam pressure at throttle, pounds,	150	175
Indicated horse power at best efficiency,	7,500	7,500
Diameter of low-pressure piston rods, inches,	8	10
Diameter of high-pressure piston rods, inches,	8	10

Diameter of crank pin, inches,	18	20
Length of crank pin, inches,	18	18
Type of Low-Pressure Valves.	Double Ported Corliss	Single Ported Corliss
Type of High-Pressure Valves.	Corliss	Poppet Type
Diameter of shaft in journals, inches,	34	34
Length of journals, inches,	60	60
Diameter of shaft in hub of revolving element, inches	37-1/16	37-1/16

The guarantees under which the main engines are being furnished, and which will govern their acceptance by the purchaser, are in substance as follows: First. The engine will be capable of operating continuously when indicating 11,000 horse power with 175 lbs. of steam pressure, a speed of 75 revolutions and a 26-inch vacuum without normal wear, jar, noise, or other objectionable results. Second. It will be suitably proportioned to withstand in a serviceable manner all sudden fluctuations of load as are usual and incidental to the generation of electrical energy for railway purposes. Third. It will be capable of operating with an atmospheric exhaust with two pounds back pressure at the low pressure cylinders, and when so operating, will fulfill all the operating requirements, except as to economy and capacity. Fourth. It will be proportioned so that when occasion shall require it can be operated with a steam pressure at the throttles of 200 pounds above atmospheric pressure under the before mentioned conditions of the speed and vacuum. Fifth. It will be proportioned so that it can be operated with steam pressure at the throttle of 200 pounds above atmospheric pressure under the before mentioned condition as to speed when exhausting in the atmosphere. Sixth. The engine will operate successfully with a steam pressure at the throttle of 175 pounds above atmosphere, should the temperature of the steam be maintained at the throttle at from 450 to 500 degrees Fahr. Seventh. It will not require more than 12-1/4 pounds of dry steam per indicated horse power per hour, when indicating 7,500 horse power at 75 revolutions per minute, when the vacuum of 26 inches at the low pressure cylinders, with a steam pressure at the throttle of 175 pounds and with saturated steam at the normal temperature due to its pressure. The guarantee includes all of the steam used by the engine or by the jackets or reheater.

The new features contained within the engine construction are principally: First, the novel construction of the high-pressure cylinders, by which only a small strain is transmitted through the valve chamber between the cylinder and the slide-surface casting. This is accomplished by employing heavy bolts, which bolt the shell of the cylinder casting to the slide-surface casting, said bolts being carried past and outside the valve chamber. Second, the use of poppet valves, which are operated in a very simple manner from a wrist plate on the side of the cylinder, the connections from the valves to the wrist plate and the connections from the wrist plate to the eccentric being similar to the parts usually employed for the operation of Corliss valves.

Unlike the Manhattan engines, the main steam pipes are carried to the high-pressure cylinders under the floor and not above it. Another modification consists in the use of an adjustable strap for the crank-pin boxes instead of the marine style of construction at the crank-pin end of the connecting rod.

The weight of the revolving field is about 335,000 pounds, which gives a flywheel effect of about 350,000 pounds at a radius of gyration of 11 feet, and with this flywheel inertia the engine is designed so that any point on the revolving element shall not, in operation, lag behind nor forge ahead of the position that it would have if the speed were absolutely uniform, by an amount greater than one-eighth of a natural degree.

Turbo-Generators

Arrangements have been made for the erection of four turbo-generators, but only three have been ordered. They are of the multiple expansion parallel flow type, consisting of two turbines arranged tandem compound. When operating at full load each of the two turbines, comprising one unit, will develop approximately equal power for direct connection to an alternator giving 7,200 alternations per minute at 11,000 volts and at a speed of 1,200 revolutions per minute. Each unit will have a normal output of 1,700 electrical horse power with a steam pressure of 175 pounds at the throttle and a vacuum in the exhaust pipe of 27 inches, measured by a mercury column and referred to a barometric pressure of 30 inches. The turbine is guaranteed to operate satisfactorily with steam superheated to 450 degrees Fahrenheit. The economy guaranteed under the foregoing conditions as to initial and terminal pressure and speed is as follows: Full load of 1,250 kilowatts, 15.7 pounds of steam per electrical horse-power hour; three-quarter load, 937-1/2 kilowatts, 16.6 pounds per electrical horse-power hour; one-half load, 625 kilowatts, 18.3 pounds; and one-quarter load, 312-1/2 kilowatts, 23.2 pounds. When operating under the conditions of speed and steam pressure mentioned, but with a pressure in the exhaust pipe of 27 inches vacuum by mercury column (referred to 30 inches barometer), and with steam at the throttle superheated 75 degrees Fahrenheit above the temperature of saturated steam at that pressure, the guaranteed steam consumption is as follows: Full load, 1,250 kilowatts, 13.8 pounds per electrical horse-power hour; three-quarter load, 937-1/2 kilowatts, 14.6 pounds; one-half load, 625 kilowatts, 16.2 pounds; and one-quarter load, 312-1/2 kilowatts, 20.8 pounds.

Exciter Engines

The two exciter engines are each direct connected to a 250 kilowatt direct current generator. Each engine is a vertical quarter-crank compound engine with a 17-inch high pressure cylinder and a 27-inch low-pressure cylinder with a common 24-inch stroke. The engines will be non-condensing, for the reason that extreme reliability is desired at the expense of some economy. They will operate at best efficiency when indicating 400 horse power at a speed of 150 revolutions per minute with a steam pressure of 175 pounds at the throttle. Each engine will have a maximum of 600 indicated horse power.

Condensing Equipment

Each engine unit is supplied with its own condenser equipment, consisting of two barometric condensing chambers, each attached as closely as possible to its respective low-pressure cylinder. For each engine also is provided a vertical circulating pump along with a vacuum pump and, for the sake of flexibility, the pumps are cross connected with those of other engines and can be used interchangeably.

The circulating pumps are vertical, cross compound pumping engines with outside packed plungers. Their foundations are upon the basement floor level and the steam cylinders extend above the engine-room floor; the starting valves and control of speed is therefore entirely under the supervision of the engineer. Each pump has a normal capacity of 10,000,000 gallons of water per day, so that the total pumping capacity of all the pumps is 120,000,000 gallons per day. While the head against which these pumps will be required to work, when assisted by the vacuum in the condenser, is much less than the total lift from low tide water to the entrance into the condensing chambers, they are so designed as to be ready to deliver the full quantity the full height, if for any reason the assistance of the vacuum should be lost or not available at times of starting up. A temporary overload can but reduce the vacuum only for a short time and the fluctuations of the tide, or even a complete loss of vacuum cannot interfere with the constant supply of water, the governor simply admitting to the cylinders the proper amount of steam to do the work. The high-pressure steam cylinder is 10 inches in diameter and the low-pressure is 20 inches; the two double-acting water plungers are each 20 inches in diameter, and the stroke is 30 inches for all. The water ends are composition fitted for salt water and have valve decks and plungers entirely of that material.

COAL UNLOADING TOWER ON WEST 58TH STREET PIER

The dry vacuum pumps are of the vertical form, and each is located alongside of the corresponding circulating pump. The steam cylinders also project above the engine-room floor. The vacuum cylinder is immediately below the steam cylinder and has a valve that is mechanically operated by an eccentric on the shaft. These pumps are of the close-clearance type, and, while controlled by a governor, can be changed in speed while running to any determined rate.

Exhaust Piping

From each atmospheric exhaust valve, which is direct-connected to the condensing chamber at each low-pressure cylinder, is run downward a 30-inch riveted-steel exhaust pipe. At a point just under the engine-room floor the exhaust pipe is carried horizontally around the engine foundations, the two from each pair of engines uniting in a 40-inch riser to the roof. This riser is between the pair of engines and back of the high-pressure cylinder, thus passing through the so-called pipe area, where it also receives exhaust steam from the pump auxiliaries. At the roof the 40-inch riser is run into a 48-inch stand pipe. This is capped with an exhaust head, the top of which is 35 feet above the roof.

All the exhaust piping 30 inches in diameter and over is longitudinally riveted steel with cast-iron flanges riveted on to it. Expansion joints are provided where necessary to relieve the piping from the strains due to expansion and contraction, and where the joints are located near the engine and generator they are of corrugated copper. The expansion joints in the 40-inch risers above the pipe area are ordinarily packed slip joints.

The exhaust piping from the auxiliaries is carried directly up into the pipe area, where it is connected with a feed-water heater, with means for by-passing the latter. Beyond the heater it joins the 40-inch riser to the roof. The feed-water heaters are three-pass, vertical, water-tube heaters, designed for a working water pressure of 225 pounds per square inch.

The design of the atmospheric relief valve received special consideration. A lever is provided to assist the valve to close, while a dash pot prevents a too quick action in either direction.

Compressed Air

The power house will be provided with a system for supplying compressed air to various points about the structure for cleaning electrical machinery and for such other purposes as may arise. It will also be used for operating whistles employed for signaling. The air is supplied to reservoir tanks by two vertical, two-stage, electric-driven air compressors.

Oil System

For the lubrication of the engines an extensive oil distributing and filtering system is provided. Filtered oil will be supplied under pressure from elevated storage tanks, with a piping system leading to all the various journals. The piping to the engines is constructed on a duplicate, or crib, system, by which the supply of oil cannot be interrupted by a break in any one pipe. The oil on leaving the engines is conducted to the filtering tanks. A pumping equipment then redelivers the oil to the elevated storage tanks.

All piping carrying filtered oil is of brass and fittings are inserted at proper pipes to facilitate cleaning. The immediate installation includes two oil filtering tanks at the easterly end of the power house, but the completed plant contemplates the addition of two extra filtering tanks at the westerly end of the structure.

Cranes, Shops, Etc.

The power house is provided with the following traveling cranes: For the operating room: One 60-ton electric traveling crane and one 25-ton electric traveling crane. For the area over the oil switches: one 10-ton hand-operated crane. For the center aisle of the boiler room: one 10-ton hand-operated crane. The span of both of the electric cranes is 74 feet 4 inches and both cranes operate over the entire length of the structure.

The 60-ton crane has two trolleys, each with a lifting capacity, for regular load, of 50 tons. Each trolley is also provided with an auxiliary hoist of 10 tons capacity. When loaded, the crane can operate at the following speeds: Bridge, 200 feet per minute; trolley, 100 feet per minute; main hoist, 10 feet per minute; and auxiliary hoist, 30 feet per minute. The 25-ton crane is provided with one trolley, having a lifting capacity, for regular load, of 25 tons, together with auxiliary hoist of 5 tons. When loaded, the crane can operate at the following speeds: bridge, 250 feet per minute; trolley, 100 feet per minute; main hoist, 12 feet per minute; and auxiliary hoist, 28 feet per minute.

The power house is provided with an extensive tool equipment for a repair and machine shop, which is located on the main gallery at the northerly side of the operating room.

5,000 K. W. ALTERNATOR—MAIN POWER HOUSE

CHAPTER V

SYSTEM OF ELECTRICAL SUPPLY

Energy from Engine Shaft to Third Rail

The system of electrical supply chosen for the subway comprises alternating current generation and distribution, and direct current operation of car motors. Four years ago, when the engineering plans were under consideration, the single-phase alternating current railway motor was not even in an embryonic state, and notwithstanding the marked progress recently made in its development, it can scarcely yet be considered to have reached a stage that would warrant any modifications in the plans adopted, even were such modifications easily possible at the present time. The comparatively limited headroom available in the subway prohibited the use of an overhead system of conductors, and this limitation, in conjunction with the obvious desirability of providing a system permitting interchangeable operation with the lines of the Manhattan Railway system practically excluded tri-phase traction systems and led directly to the adoption of the third-rail direct current system.

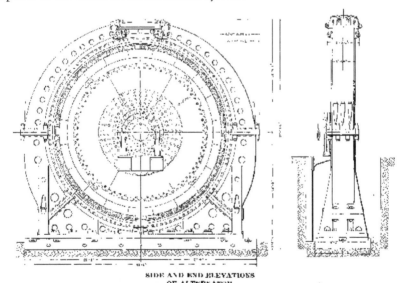

SIDE AND END ELEVATIONS OF ALTERNATOR.

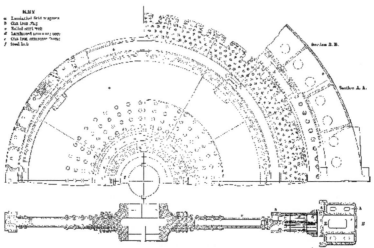

KEY
a Laminated field w spools
b Coil iron ring
c Radial steel web
d Laminated armature core
e Coil iron armature frame
f Steel bolt

SIDE ELEVATION AND CROSS SECTION OF ALTERNATOR WITH PART CUT AWAY TO SHOW CONSTRUCTION.

It being considered impracticable to predict with entire certainty the ultimate traffic conditions to be met, the generator plant has been designed to take care of all probable traffic demands expected to arise within a year or two of the beginning of operation of the system, while the plans permit convenient and symmetrical increase to meet the requirements of additional demand which may develop. Each express train will comprise five motor cars and three trail cars, and each local train will comprise three motor cars and two trail cars. The weight of each motor car with maximum live load is 88,000 pounds, and the weight of each trailer car 66,000 pounds.

The plans adopted provide electric equipment at the outstart capable of operating express trains at an average speed approximating twenty-five miles per hour, while the control system and motor units have been so chosen that higher speeds up to a limit of about thirty miles per hour can be attained by increasing the number of motor cars providing experience in operation demonstrates that such higher speeds can be obtained with safety.

The speed of local trains between City Hall and 96th Street will average about 15 miles an hour, while north of 96th Street on both the West side and East side branches their speed will average about 18 miles an hour, owing to the greater average distance between local stations.

As the result of careful consideration of various plans, the company's engineers recommended that all the power required for the operation of the system be generated in a single power house in the form of three-phase alternating current at 11,000 volts, this current to be generated at a frequency of 25 cycles per second, and to be delivered through three-conductor cables to transformers and converters in sub-stations suitably located with reference to the track system, the current there to be transformed and converted to direct current for delivery to the third-rail conductor at a potential of 625 volts.

OPERATING GALLERY IN SUB-STATION

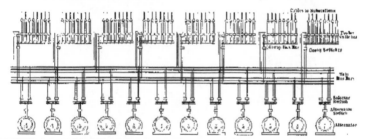

GENERAL DIAGRAM OF 11,000 VOLT CIRCUITS IN MAIN POWER STATION

Calculations based upon contemplated schedules require for traction purposes and for heating and lighting cars, a maximum delivery of about 45,000 kilowatts at the third rail. Allowing for losses in the distributing cables, in transformers and converters, this implies a total generating capacity of approximately 50,000 kilowatts, and having in view the possibility of future extensions of the system it was decided to design and construct the power house building for the ultimate reception of eleven 5,000-kilowatt units for traction current in addition to the lighting sets. Each 5,000-kilowatt unit is capable of delivering during rush hours an output of 7,500 kilowatts or approximately 10,000 electrical horse power and, setting aside one unit as a reserve, the contemplated ultimate maximum output of the power plant, therefore, is 75,000 kilowatts, or approximately 100,000 electrical horse power.

Power House

The power house is fully described elsewhere in this publication, but it is not inappropriate to refer briefly in this place to certain considerations governing the selection of the generating unit, and the use of engines rather than steam turbines.

OIL SWITCHES—MAIN POWER STATION

The 5,000-kilowatt generating unit was chosen because it is practically as large a unit of the direct-connected type as can be constructed by the engine builders unless more than two bearings be used—an alternative deemed inadvisable by the engineers of the company. The adoption of a smaller unit would be less economical of floor space and would tend to produce extreme complication in so large an installation, and, in view of the rapid changes in load which in urban railway service of this character occur in the morning and again late in the afternoon, would be extremely difficult to operate.

The experience of the Manhattan plant has shown, as was anticipated in the installation of less output than this, the alternators must be put in service at intervals of twenty minutes to meet the load upon the station while it is rising to the maximum attained during rush hours.

After careful consideration of the possible use of steam turbines as prime-movers to drive the alternators, the company's engineers decided in favor of reciprocating engines. This decision was made three years ago and, while the steam turbine since that time has made material progress, those responsible for the decision are confirmed in their opinion that it was wise.

PART OF BUS BAR COMPARTMENTS—MAIN POWER STATION

Alternators

The alternators closely resemble those installed by the Manhattan Railway Company (now the Manhattan division of the Interborough Rapid Transit Company) in its plant on the East River, between 74th Street and 75th Street. They differ, however, in having the stationary armature divided into seven castings instead of six, and in respect to details of the armature winding. They are three-phase machines, delivering twenty-five cycle alternating currents at an effective potential of 11,000 volts. They are 42 feet in height, the diameter of the revolving part is 32 feet, its weight, 332,000 pounds, and the aggregate weight of the machine, 889,000 pounds. The design of the engine dynamo unit eliminates the auxiliary fly wheel generally used in the construction of large direct-connected units prior to the erection of the Manhattan plant, the weight and dimensions of the revolving alternator field being such with reference to the turning moment of the engine as to secure close uniformity of rotation, while at the same time this construction results in narrowing the engine and reducing the engine shafts between bearings.

REAR VIEW OF BUS BAR COMPARTMENTS—MAIN POWER STATION

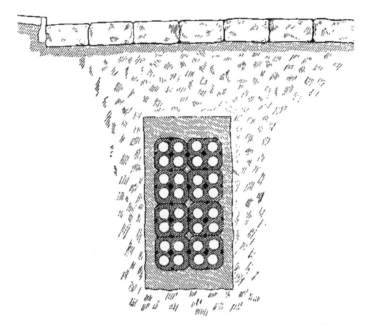

DUCT LINE ACROSS 58TH STREET 32 DUCTS

Construction of the revolving parts of the alternators is such as to secure
very great strength and consequent ability to resist the tendency to burst
and fly apart in case of temporary abnormal speed through accident of any
kind. The hub of the revolving field is of cast steel, and the rim is carried
not by the usual spokes but by two wedges of rolled steel. The construction
of the revolving field is illustrated on pages 91 and 92. The angular velocity
of the revolving field is remarkably uniform. This result is due primarily to
the fact that the turning movement of the four-cylinder engine is far more
uniform than is the case, for example, with an ordinary two-cylinder engine.
The large fly-wheel capacity of the rotating element of the machine also
contributes materially to secure uniformity of rotation.

MAIN CONTROLLING BOARD IN POWER STATION

CONTROL AND INSTRUMENT BOARD—MAIN POWER
STATION

The alternators have forty field poles and operates at seventy-five revolutions per minute. The field magnets constitute the periphery of the revolving field, the poles and rim of the field being built up by steel plates which are dovetailed to the driving spider. The heavy steel end plates are bolted together, the laminations breaking joints in the middle of the pole. The field coils are secured by copper wedges, which are subjected to shearing strains only. In the body of the poles, at intervals of approximately three inches, ventilating spaces are provided, these spaces registering with corresponding air ducts in the external armature. The field winding consists of copper strap on edge, one layer deep, with fibrous material cemented in place between turns, the edges of the strap being exposed.

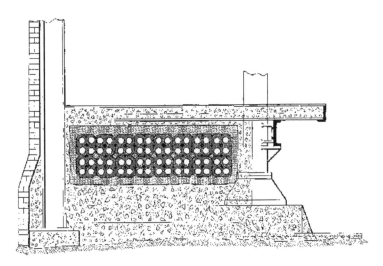

DUCTS UNDER PASSENGER STATION PLATFORM 64
DUCTS

The armature is stationary and exterior to the field. It consists of a laminated ring with slots on its inner surface and supported by a massive external cast-iron frame. The armature, as has been noted, comprises seven segments, the topmost segment being in the form of a small keystone. This may be removed readily, affording access to any field coil, which in this way may be easily removed and replaced. The armature winding consists of U-shaped copper bars in partially closed slots. There are four bars per slot and three slots per phase per pole. The bars in any slot may be removed from the armature without removing the frame. The alternators, of course, are separately excited, the potential of the exciting current used being 250 volts.

As regards regulation, the manufacturer's guarantee is that at 100 per cent. power factor if full rated load be thrown off the e. m. f. will rise 6 per cent. with constant speed and constant excitation. The guarantee as to efficiency is as follows: On non-inductive load, the alternators will have an efficiency of not less than 90.5 per cent. at one-quarter load; 94.75 per cent. at one-half load; 96.25 per cent. at three-quarters load; 97 per cent. at full load, and 97.25 per cent. at one and one-quarter load. These figures refer, of course, to electrical efficiency, and do not include windage and bearing friction. The machines are designed to operate under their rated full load with rise of temperature not exceeding 35 degrees C. after twenty-four hours.

THREE-CONDUCTOR NO. 000 CABLE FOR 11,000 VOLT
DISTRIBUTION

Exciters

To supply exciting current for the fields of the alternators and to operate motors driving auxiliary apparatus, five 250-kilowatt direct current dynamos are provided. These deliver their current at a potential of 250 volts. Two of them are driven by 400 horse-power engines of the marine type, to which they are direct-connected, while the remaining three units are direct-connected to 365 horse-power tri-phase induction motors operating at 400 volts. A storage battery capable of furnishing 3,000 amperes for one hour is used in co-operation with the dynamos provided to excite the alternators. The five direct-current dynamos are connected to the organization of switching apparatus in such a way that each unit may be connected at will either to the exciting circuits or to the circuits through which auxiliary motors are supplied.

The alternators for which the new Interborough Power House are designed will deliver to the bus bars 100,000 electrical horse power. The current delivered by these alternators reverses its direction fifty times per second and in connecting dynamos just coming into service with those already in operation the allowable difference in phase relation at the instant the circuit is completed is, of course, but a fraction of the fiftieth of a second. Where the power to be controlled is so great, the potential so high, and the speed requirements in respect to synchronous operation so exacting, it is obvious that the perfection of control attained in some of our modern plants is not their least characteristic.

Switching Apparatus

The switch used for the 11,000-volt circuits is so constructed that the circuits are made and broken under oil, the switch being electrically operated. Two complete and independent sets of bus bars are used, and the connections are such that each alternator and each feeder may be connected to either of these sets of bus bars at the will of the operator. From alternators to bus bars the current passes, first, through the alternator switch, and then alternatively through one or the other of two selector switches which are connected, respectively, to the two sets of bus bars.

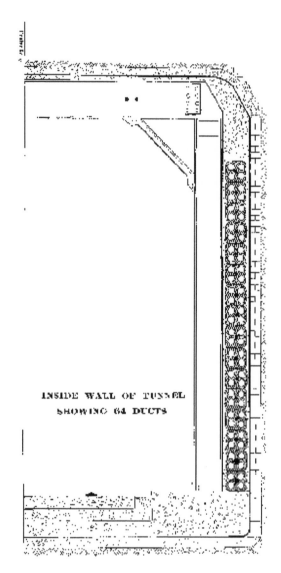

INSIDE WALL OF TUNNEL SHOWING 64 DUCTS

Provision is made for an ultimate total of twelve sub-stations, to each of which as many as eight feeders may be installed if the development of the company's business should require that number. But eight sub-stations are required at present, and to some of these not more than three feeders each are necessary. The aggregate number of feeders installed for the initial operation of the subway system is thirty-four.

Each feeder circuit is provided with a type H-oil switch arranged to be open and closed at will by the operator, and also to open automatically in the case of abnormal flow of current through the feeder. The feeders are arranged in groups, each group being supplied from a set of auxiliary bus bars, which in turn receives its supply from one or the other of the two sets of main bus bars; means for selection being provided as in the case of the alternator circuits by a pair of selector switches, in this case designated as group switches. The diagram on page 93 illustrates the essential features of the organization and connections of the 11,000-volt circuits in the power house.

MANHOLES IN SIDE WALL OF SUBWAY

Any and every switch can be opened or closed at will by the operator standing at the control board described. The alternator switches are provided also with automatic overload and reversed current relays, and the feeder switches, as above mentioned, are provided with automatic overload relays. These overload relays have a time attachment which can be set to open the switch at the expiration of a predetermined time ranging from .3 of a second to 5 seconds.

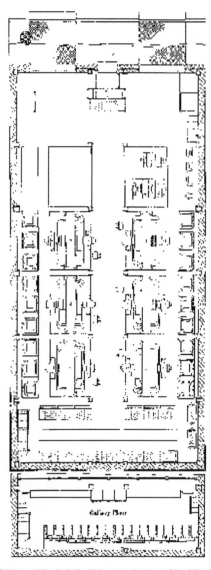

CONVERTER FLOOR PLAN SUB-STATION NO. 14

The type H-oil switch is operated by an electric motor through the intervention of a mechanism comprising powerful springs which open and close the switch with great speed. This switch when opened introduces in each of the three sides of the circuit two breaks which are in series with each other. Each side of the circuit is separated from the others by its location in an enclosed compartment, the walls of which are brick and soapstone. The general construction of the switch is illustrated by the photograph on page 94.

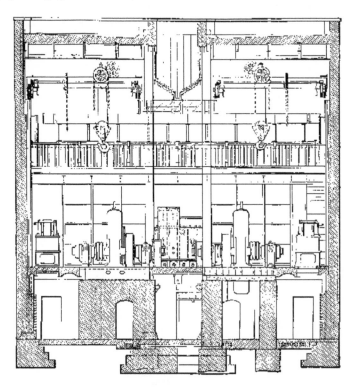

CROSS SECTION SUB-STATION NO. 14

INTERIOR OF SUB-STATION NO. 11

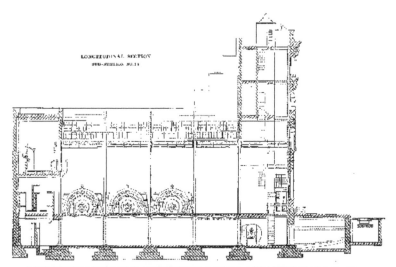

LONGITUDINAL SECTION SUB-STATION NO. 14

Like all current-carrying parts of the switches, the bus bars are enclosed in separate compartments. These are constructed of brick, small doors for inspection and maintenance being provided opposite all points where the bus bars are supported upon insulators. The photographs on pages 95 and 96 are views of a part of the bus bar and switch compartments.

TWO GROUPS OF TRANSFORMERS

The oil switches and group bus bars are located upon the main floor and extend along the 59th Street wall of the engine room a distance of about 600 feet. The main bus bars are arranged in two lines of brick compartments, which are placed below the engine-room floor. These bus bars are arranged vertically and are placed directly beneath the rows of oil switches located upon the main floor of the power house. Above these rows of oil switches and the group bus bars, galleries are constructed which extend the entire length of the power house, and upon the first of these galleries at a point opposite the middle of the power house are located the control board and instrument board, by means of which the operator in charge regulates and directs the entire output of the plant, maintaining a supply of power at all times adequate to the demands of the transportation service.

**MOTOR-GENERATORS AND BATTERY BOARD FOR
CONTROL CIRCUITS—SUB-STATION**

1,500 K. W. ROTARY CONVERTER

The control board is shown in the photograph on page 97. Every alternator switch, every selector switch, every group switch, and every feeder switch upon the main floor is here represented by a small switch. The small switch is connected into a control circuit which receives its supply of energy at 110 volts from a small motor generator set and storage battery. The motors which actuate the large oil switches upon the main floor are driven by this 110 volt control current, and thus in the hands of the operator the control switches make or break the relatively feeble control currents, which, in turn, close or open the switches in the main power circuits. The control switches are systematically assembled upon the control bench board in conjunction with dummy bus bars and other apparent (but not real) metallic connections, the whole constituting at all times a correct diagram of the existing connections of the main power circuits. Every time the operator changes a connection by opening or closing one of the main switches, he necessarily changes his diagram so that it represents the new conditions established by opening or closing the main switch. In connection with each control switch two small bull's-eye lamps are used, one red, to indicate that the corresponding main switch is closed, the other green, to indicate that it is open. These lamps are lighted when the moving part of the main switch reaches approximately the end of its travel. If for any reason, therefore, the movement of the control switch should fail to actuate the main switch, the indicator lamp will not be lighted.

MOTOR-GENERATOR SET SUPPLYING ALTERNATING CURRENT FOR BLOCK SIGNALS AND MOTOR-GENERATOR STARTING SET

The control board is divided into two parts—one for the connections of the alternators to the bus bars and the other for the connection of feeders to bus bars. The drawing on page 97 shows in plain view the essential features of the control boards.

The Instrument Board

A front view of the Instrument Board is shown on page 97. This board contains all indicating instruments for alternators and feeders. It also carries standardizing instruments and a clock. In the illustration the alternator panels are shown at the left and the feeder panels at the right. For the alternator panels, instruments of the vertical edgewise type are used. Each vertical row comprises the measuring instruments for an alternator. Beginning at the top and enumerating them in order these instruments are: Three ammeters, one for each phase, a volumeter, an indicating wattmeter, a power factor indicator and a field ammeter. The round dial instrument shown at the bottom of each row of instruments is a three-phase recording wattmeter.

A panel located near the center of the board between alternator panels and feeder panels carries standard instruments used for convenient calibration of the alternator and feeder instruments. Provision is made on the back of the board for convenient connection of the standard instruments in series with the instruments to be compared. The panel which carries the standard instruments also carries ammeters used to measure current to auxiliary circuits in the power house.

For the feeder board, instruments of the round dial pattern are used, and for each feeder a single instrument is provided, viz., an ammeter. Each vertical row comprises the ammeters belonging to the feeders which supply a given sub-station, and from left to right these are in order sub-stations Nos. 11, 12, 13, 14, 15, 16, 17, and 18; blank spaces are left for four additional sub-stations. Each horizontal row comprises the ammeter belonging to feeders which are supplied through a given group switch.

This arrangement in vertical and horizontal lines, indicating respectively feeders to given sub-stations and feeders connected to the several group switches, is intended to facilitate the work of the operator. A glance down a vertical row without stopping to reach the scales of the instruments will tell him whether the feeders are dividing with approximate equality the load to a given sub-station. Feeders to different sub-stations usually carry different loads and, generally speaking, a glance along a horizontal row will convey no information of especial importance. If, however, for any reason the operator should desire to know the approximate aggregate load upon a group of feeders this systematic arrangement of the instruments is of use.

**SWITCHBOARD FOR ALTERNATING CURRENT BLOCK
SIGNAL CIRCUITS—IN SUB-STATION**

EXTERIOR OF SUB-STATION NO. 18

From alternators to alternator switches the 11,000 volt alternating currents are conveyed through single conductor cables, insulated by oil cambric, the thickness of the wall being 12/32 of an inch. These conductors are installed in vitrified clay ducts. From dynamo switches to bus bars and from bus bars to group and feeder switches, vulcanized rubber insulation containing 30 per cent. pure Para rubber is employed. The thickness of insulating wall is 9/32 of an inch and the conductors are supported upon porcelain insulators.

Conduit System for Distribution

From the power house to the subway at 58th Street and Broadway two lines of conduit, each comprising thirty-two ducts, have been constructed. These conduits are located on opposite sides of the street. The arrangement of ducts is 8 x 4, as shown in the section on page 96.

EXTERIOR OF SUB-STATION NO. 11

The location and arrangement of ducts along the line of the subway are illustrated in photographs on pages 98 and 99, which show respectively a section of ducts on one side of the subway, between passenger stations, and a section of ducts and one side of the subway, beneath the platform of a passenger station. From City Hall to 96th Street (except through the Park Avenue Tunnel) sixty-four ducts are provided on each side of the subway. North of 96th Street sixty-four ducts are provided for the West-side lines and an equal number for the East-side lines. Between passenger stations these ducts help to form the side walls of the subway, and are arranged thirty-two ducts high and two ducts wide. Beneath the platform of passenger stations the arrangement is somewhat varied because of local obstructions, such as pipes, sewers, etc., of which it was necessary to take account in the construction of the stations. The plan shown on page 98, however, is typical.

The necessity of passing the cables from the 32 x 2 arrangement of ducts along the side of the tunnel to 8 x 8 and 16 x 4 arrangements of ducts beneath the passenger platforms involves serious difficulties in the proper support and protection of cables in manholes at the ends of the station platforms. In order to minimize the risk of interruption of service due to possible damage to a considerable number of cables in one of these manholes, resulting from short circuit in a single cable, all cables except at the joints are covered with two layers of asbestos aggregating a full 1/4-inch in thickness. This asbestos is specially prepared and is applied by wrapping the cable with two strips each 3 inches in width, the outer strip covering the line of junction between adjacent spirals of the inner strip, the whole when in place being impregnated with a solution of silicate of soda. The joints themselves are covered with two layers of asbestos held in place by steel tape applied spirally. To distribute the strains upon the cables in manholes, radical supports of various curvatures, and made of malleable cast iron, are used. The photograph on page 100 illustrates the arrangement of cables in one of these manholes.

OPERATING BOARD—SUB-STATION NO. 11

In order to further diminish the risk of interruption of the service due to failure of power supply, each sub-station south of 96th Street receives its alternating current from the power house through cables carried on opposite sides of the subway. To protect the lead sheaths of the cables against damage by electrolysis, rubber insulating pieces 1/6 of an inch in thickness are placed between the sheaths and the iron bracket supports in the manholes.

Cable Conveying Energy from Power House to Sub-Stations

The cables used for conveying energy from the power house to the several sub-stations aggregate approximately 150 miles in length. The cable used for this purpose comprises three stranded copper conductors each of which contains nineteen wires, and the diameter of the stranded conductor thus formed is 2/5 of an inch. Paper insulation is employed and the triple cable is enclosed in a lead sheath 9/64 of an inch thick. Each conductor is separated from its neighbors and from the lead sheath by insulation of treated paper 7/16 of an inch in thickness. The outside diameter of the cables is 2-5/8 inches, and the weight 8-1/2 pounds per lineal foot. In the factories the cable as manufactured was cut into lengths corresponding to the distance between manholes, and each length subjected to severe tests including application to the insulation of an alternating current potential of 30,000 volts for a period of thirty minutes. These cables were installed under the supervision of the Interborough Company's engineers, and after jointing, each complete cable from power house to sub-station was tested by applying an alternating potential of 30,000 volts for thirty minutes between each conductor and its neighbors, and between each conductor and the lead sheath. The photographs on page 98 illustrates the construction of this cable.

Sub-Station

The tri-phase alternating current generated at the power house is conveyed through the high potential cable system to eight sub-stations containing the necessary transforming and converting machinery. These sub-stations are designed and located as follows:

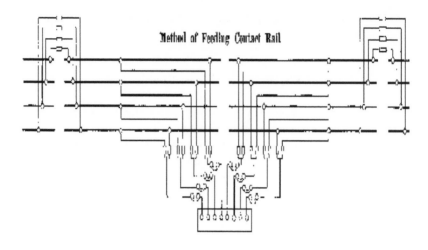

Method of Feeding Contact Rail

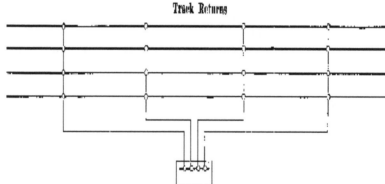

Track Returns

DIAGRAMS OF DIRECT CURRENT FEEDER AND RETURN CIRCUITS

Sub-station No. 11—29-33 City Hall Place.

Sub-station No. 12—108-110 East 19th Street.

Sub-station No. 13—225-227 West 53d Street.

Sub-station No. 14—264-266 West 96th Street.

Sub-station No. 15—606-608 West 143d Street.

Sub-station No. 16—73-77 West 132d Street.

Sub-station No. 17—Hillside Avenue, 301 feet West of Eleventh Avenue.

Sub-station No. 18—South side of Fox Street (Simpson Street), 60 feet north of Westchester Avenue.

SWITCH CONNECTING FEEDER TO CONTACT RAIL

CONTACT RAIL JOINT WITH FISH PLATE

The converter unit selected to receive the alternating current and deliver direct current to the track, etc., has an output of 1,500 kilowatts with ability to carry 50 per cent. overload for three hours. The average area of a city lot is 25 x 100 feet, and a sub-station site comprising two adjacent lots of this approximate size permits the installation of a maximum of eight 1,500 kilowatts converters with necessary transformers, switchboard and other auxiliary apparatus. In designing the sub-stations, a type of building with a central air-well was selected. The typical organization of apparatus is illustrated in the ground plan and vertical section on pages 101, 102 and 103 and provides, as shown, for two lines of converters, the three transformers which supply each converter being located between it and the adjacent side wall. The switchboard is located at the rear of the station. The central shaft affords excellent light and ventilation for the operating room. The steel work of the sub-stations is designed with a view to the addition of two storage battery floors, should it be decided at some future time that the addition of such an auxiliary is advisable.

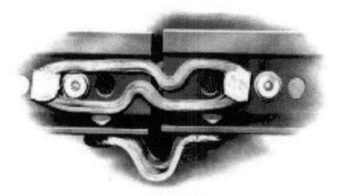

CONTACT RAIL BANDS

The necessary equipment of the sub-stations implies sites approximately 50 x 100 feet in dimensions; and sub-stations Nos. 14, 15, 17, and 18 are practically all this size. Sub-stations Nos. 11 and 16 are 100 feet in length, but the lots acquired in these instances being of unusual width, these sub-stations are approximately 60 feet wide. Sub-station No. 12, on account of limited ground space, is but 48 feet wide and 92 feet long. In each of the sub-stations, except No. 13, foundations are provided for eight converters; sub-station No. 13 contains foundations for the ultimate installation of ten converters.

DIRECT CURRENT FEEDERS FROM MANHOLE TO
CONTACT RAIL

The function of the electrical apparatus in sub-stations, as has been stated, is the conversion of the high potential alternating current energy delivered from the power house through the tri-phase cables into direct current adapted to operate the motors with which the rolling stock is equipped. This apparatus comprises transformers, converters, and certain minor auxiliaries. The transformers, which are arranged in groups of three, receive the tri-phase alternating current at a potential approximating 10,500 volts, and deliver equivalent energy (less the loss of about 2 per cent. in the transformation) to the converters at a potential of about 390 volts. The converters receiving this energy from their respective groups of transformers in turn deliver it (less a loss approximating 4 per cent. at full load) in the form of direct current at a potential of 625 volts to the bus bars of the direct current switchboards, from which it is conveyed by insulated cables to the contact rails. The photograph on page 102 is a general view of the interior of one of the sub-stations. The exterior of sub-stations Nos. 11 and 18 are shown on page 107.

CONTACT RAILS, SHOWING END INCLINES

The illustration on page 108 is from a photograph taken on one of the switchboard galleries. In the sub-stations, as in the power house, the high potential alternating current circuits are opened and closed by oil switches, which are electrically operated by motors, these in turn being controlled by 110 volt direct current circuits. Diagramatic bench boards are used, as at the power house, but in the sub-stations they are of course relatively small and free from complication.

The instrument board is supported by iron columns and is carried at a sufficient height above the bench board to enable the operator, while facing the bench board and the instruments, to look out over the floor of the sub-station without turning his head. The switches of the direct current circuits are hand-operated and are located upon boards at the right and left of the control board.

A novel and important feature introduced (it is believed for the first time) in these sub-stations, is the location in separate brick compartments of the automatic circuit breakers in the direct current feeder circuits. These circuit breaker compartments are shown in the photograph on page 93, and are in a line facing the boards which carry the direct feeder switches, each circuit breaker being located in a compartment directly opposite the panel which carries the switch belonging to the corresponding circuit. This plan will effectually prevent damage to other parts of the switchboard equipment when circuit-breakers open automatically under conditions of short-circuit. It also tends to eliminate risk to the operator, and, therefore, to increase his confidence and accuracy in manipulating the hand-operated switches.

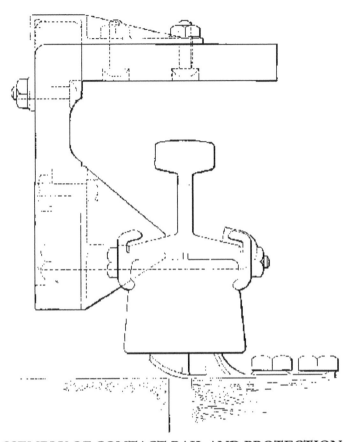

ASSEMBLY OF CONTACT RAIL AND PROTECTION

The three conductor cables which convey tri-phase currents from the power house are carried through tile ducts from the manholes located in the street directly in front of each sub-station to the back of the station where the end of the cable is connected directly beneath its oil switch. The three conductors, now well separated, extend vertically to the fixed terminals of the switch. In each sub-station but one set of high-potential alternating current bus bars is installed and between each incoming cable and these bus bars is connected an oil switch. In like manner, between each converter unit and the bus bars an oil switch is connected into the high potential circuit. The bus bars are so arranged that they may be divided into any number of sections not exceeding the number of converter units, by means of movable links which, in their normal condition, constitute a part of the bus bars.

Each of the oil switches between incoming circuits and bus bars is arranged for automatic operation and is equipped with a reversed current relay, which, in the case of a short-circuit in its alternating current feeder cable opens the switch and so disconnects the cable from the sub-station without interference with the operation of the other cables or the converting machinery.

CONTACT RAIL INSULATOR

Direct Current Distribution from Sub-Stations

The organization of electrical conductors provided to convey direct current from the sub-stations to the moving trains can be described most conveniently by beginning with the contact, or so-called third rail. South of 96th Street the average distance between sub-stations approximates 12,000 feet, and north of 96th Street the average distance is about 15,000 feet. Each track, of course, is provided with a contact rail. There are four tracks and consequently four contact rails from City Hall to 96th Street, three from 96th Street to 145th Street on the West Side, two from 145th Street to Dyckman Street, and three from Dyckman Street to the northern terminal of the West Side extension of the system. From 96th Street, the East Side has two tracks and two contact rails to Mott Avenue, and from that point to the terminal at 182d Street three tracks and three contact rails.

CONTACT SHOE AND FUSE

Contact rails south of Reade Street are supplied from sub-station No. 11; from Reade Street to 19th Street they are supplied from sub-stations Nos. 11 and 12; from 19th Street they are supplied from sub-stations Nos. 12 and 13; from the point last named to 96th Street they are supplied from sub-stations Nos. 13 and 14; from 96th Street to 143d Street, on the West Side, they are supplied from sub-stations Nos. 14 and 15; from 143d Street to Dyckman Street they are supplied from sub-stations Nos. 15 and 17; and from that point to the terminal they are supplied from sub-station No. 17. On the East Side branch contact rails from 96th Street to 132d Street are supplied from sub-stations Nos. 14 and 16; from 132d to 165th Street they are supplied from sub-stations Nos. 16 and 18; and from 165th Street to 182d Street they are supplied from sub-station No. 18.

Each contact rail is insulated from all contact rails belonging to adjacent tracks. This is done in order that in case of derailment or other accident necessitating interruption of service on a given track, trains may be operated upon the other tracks having their separate and independent channels of electrical supply. To make this clear, we may consider that section of the subway which lies between Reade Street and 19th Street. This section is equipped with four tracks, and the contact rail for each track, together with the direct current feeders which supply it from sub-stations Nos. 11 and 12, are electrically insulated from all other circuits. Of each pair of track rails one is used for the automatic block signaling system, and, therefore, is not used as a part of the negative or return side of the direct current system. The other four track rails, however, are bonded, and together with the negative feeders constitute the track return or negative side of the direct current system.

The diagram on page 109 illustrates the connections of the contact rails, track rails and the positive and negative feeders. All negative as well as positive feeders are cables of 2,000,000 c. m. section and lead sheathed. In emergency, as, for example, in the case of the destruction of a number of the cables in a manhole, they are, therefore, interchangeable. The connections are such as to minimize "track drop," as will be evident upon examination of the diagram. The electrical separation of the several contact rails and the positive feeders connected thereto secures a further important advantage in permitting the use at sub-stations of direct-current circuit-breakers of moderate size and capacity, which can be set to open automatically at much lower currents than would be practicable were all contact rails electrically connected, thus reducing the limiting current and consequently the intensity of the arcs which might occur in the subway in case of short-circuit between contact rail and earth.

The contact rail itself is of special soft steel, to secure high conductivity. Its composition, as shown by tests, is as follows: Carbon, .08 to .15; silicon, .05; phosphorus, .10; manganese, .50 to .70; and sulphur, .05. Its resistance is not more than eight times the resistance of pure copper of equal cross-section. The section chosen weighs 75 pounds per yard. The length used in general is 60 feet, but in some cases 40 feet lengths are substituted. The contact rails are bounded by four bonds, aggregating 1,200,000 c. m. section. The bonds are of flexible copper and their terminals are riveted to the steel by hydraulic presses, producing a pressure of 35 tons. The bonds when in use are covered by special malleable iron fish-plates which insure alignment of rail. Each length of rail is anchored at its middle point and a small clearance is allowed between ends of adjacent rails for expansion and contraction, which in the subway, owing to the relatively small change of temperature, will be reduced to a minimum. The photographs on pages 110 and 111 illustrate the method of bonding the rail, and show the bonded joint completed by the addition of the fish-plates.

The contact rail is carried upon block insulators supported upon malleable iron castings. Castings of the same material are used to secure the contact rail in position upon the insulators. A photograph of the insulator with its castings is shown on page 113.

The track rails are 33 feet long, of Standard American Society Civil Engineers' section, weighing 100 pounds a yard. As has been stated, one rail in each track is used for signal purposes and the other is utilized as a part of the negative return of the power system. Adjacent rails to be used for the latter purpose are bonded with two copper bonds having an aggregate section of 400,000 c. m. These bonds are firmly riveted into the web of the rail by screw bonding presses. They are covered by splice bars, designed to leave sufficient clearance for the bond.

The return rails are cross-sectioned at frequent intervals for the purpose of equalizing currents which traverse them.

Contact Rail Guard and Collector Shoe

The Interborough Company has provided a guard in the form of a plank 8-1/2 inches wide and 1-1/2 inches thick, which is supported in a horizontal position directly above the rail, as shown in the illustration on page 113. This guard is carried by the contact rail to which it is secured by supports, the construction of which is sufficiently shown in the illustration. This type of guard has been in successful use upon the Wilkesbarre and Hazleton Railway for nearly two years. It practically eliminates the danger from the third rail, even should passengers leave the trains and walk through a section of the tunnel while the rails are charged.

Its adoption necessitates the use of a collecting shoe differing radically from that used upon the Manhattan division and upon the elevated railways employing the third rail system in Chicago, Boston, Brooklyn, and elsewhere. The shoe is shown in the photograph on page 114. The shoe is held in contact with the third rail by gravity reinforced by pressure from two spiral springs. The support for the shoe includes provision for vertical adjustment to compensate for wear of car wheels, etc.

CHAPTER VI

ELECTRICAL EQUIPMENT OF CARS

In determining the electrical equipment of the trains, the company has aimed to secure an organization of motors and control apparatus easily adequate to operate trains in both local and express service at the highest speeds compatible with safety to the traveling public. For each of the two classes of service the limiting safe speed is fixed by the distance between stations at which the trains stop, by curves, and by grades. Except in a few places, for example where the East Side branch passes under the Harlem River, the tracks are so nearly level that the consideration of grade does not materially affect determination of the limiting speed. While the majority of the curves are of large radius, the safe limiting speed, particularly for the express service, is necessarily considerably less than it would be on straight tracks.

The average speed of express trains between City Hall and 145th Street on the West Side will approximate 25 miles an hour, including stops. The maximum speed of trains will be 45 miles per hour. The average speed of local and express trains will exceed the speed made by the trains on any elevated railroad.

To attain these speeds without exceeding maximum safe limiting speeds between stops, the equipment provided will accelerate trains carrying maximum load at a rate of 1.25 miles per hour per second in starting from stations on level track. To obtain the same acceleration by locomotives, a draw-bar pull of 44,000 pounds would be necessary—a pull equivalent to the maximum effect of six steam locomotives such as were used recently upon the Manhattan Elevated Railway in New York, and equivalent to the pull which can be exerted by two passenger locomotives of the latest Pennsylvania Railroad type. Two of these latter would weigh about 250 net tons. By the use of the multiple unit system of electrical control, equivalent results in respect to rate of acceleration and speed are attained, the total addition to train weight aggregating but 55 net tons.

If the locomotive principle of train operation were adopted, therefore, it is obvious that it would be necessary to employ a lower rate of acceleration for express trains. This could be attained without very material sacrifice of average speed, since the average distance between express stations is nearly two miles. In the case of local trains, however, which average nearly three stops per mile, no considerable reduction in the acceleration is possible without a material reduction in average speed. The weight of a local train exceeds the weight of five trail cars, similarly loaded, by 33 net tons, and equivalent adhesion and acceleration would require locomotives having not less than 80 net tons effective upon drivers.

Switching

The multiple unit system adopted possesses material advantages over a locomotive system in respect to switching at terminals. Some of the express trains in rush hours will comprise eight cars, but at certain times during the day and night when the number of people requiring transportation is less than during the morning and evening, and were locomotives used an enormous amount of switching, coupling and uncoupling would be involved by the comparative frequent changes of train lengths. In an eight-car multiple-unit express train, the first, third, fifth, sixth, and eighth cars will be motor cars, while the second, fourth, and seventh will be trail cars. An eight-car train can be reduced, therefore, to a six-car train by uncoupling two cars from either end, to a five-car train by uncoupling three cars from the rear end, or to a three-car train by uncoupling five cars from either end. In each case a motor car will remain at each end of the reduced train. In like manner, a five-car local train may be reduced to three cars, still leaving a motor car at each end by uncoupling two cars from either end, since in the normal five-car local train the first, third, and fifth cars will be motor cars.

200 H. P. RAILWAY MOTOR

Motors

The motors are of the direct current series type and are rated 200 horse power each. They have been especially designed for the subway service in line with specifications prepared by engineers of the Interborough Company, and will operate at an average effective potential of 570 volts. They are supplied by two manufacturers and differ in respect to important features of design and construction, but both are believed to be thoroughly adequate for the intended service.

200 H. P. RAILWAY MOTOR

The photographs on this page illustrate motors of each make. The weight of one make complete, with gear and gear case, is 5,900 pounds. The corresponding weight of the other is 5,750 pounds. The ratio of gear reduction used with one motor is 19 to 63, and with the other motor 20 to 63.

200 H. P. RAILWAY MOTOR

By the system of motor control adopted for the trains, the power delivered to the various motors throughout the train is simultaneously controlled and regulated by the motorman at the head of the train. This is accomplished by means of a system of electric circuits comprising essentially a small drum controller and an organization of actuating circuits conveying small currents which energize electric magnets placed beneath the cars, and so open and close the main power circuits which supply energy to the motors. A controller is mounted upon the platform at each end of each motor car, and the entire train may be operated from any one of the points, the motorman normally taking his post on the front platform of the first car. The switches which open and close the power circuits through motors and rheostats are called contactors, each comprising a magnetic blow-out switch and the electro magnet which controls the movements of the switch. By these contactors the usual series-multiple control of direct-current motors is effected. The primary or control circuits regulate the movement, not only of the contactors but also of the reverser, by means of which the direction of the current supplied to motors may be reversed at the will of the motorman.

APPARATUS UNDER COMPOSITE MOTOR CAR

The photograph on this page shows the complete control wiring and motor equipment of a motor car as seen beneath the car. In wiring the cars unusual precautions have been adopted to guard against risk of fire. As elsewhere described in this publication, the floors of all motor cars are protected by sheet steel and a material composed of asbestos and silicate of soda, which possesses great heat-resisting properties. In addition to this, all of the important power wires beneath the car are placed in conduits of fireproof material, of which asbestos is the principal constituent. Furthermore, the vulcanized rubber insulation of the wires themselves is covered with a special braid of asbestos, and in order to diminish the amount of combustible insulating material, the highest grade of vulcanized rubber has been used, and the thickness of the insulation correspondingly reduced. It is confidently believed that the woodwork of the car body proper cannot be seriously endangered by an accident to the electric apparatus beneath the car. Insulation is necessarily combustible, and in burning evolves much smoke; occasional accidents to the apparatus, notwithstanding every possible precaution, will sometimes happen; and in the subway the flash even of an absolutely insignificant fuse may be clearly visible and cause alarm. The public traveling in the subway should remember that even very severe short-circuits and extremely bright flashes beneath the car involve absolutely no danger to passengers who remain inside the car.

The photograph on page 120 illustrates the control wiring of the new steel motorcars. The method of assembling the apparatus differs materially from that adopted in wiring the outfit of cars first ordered, and, as the result of greater compactness which has been attained, the aggregate length of the wiring has been reduced one-third.

The quality and thickness of the insulation is the same as in the case of the earlier cars, but the use of asbestos conduits is abandoned and iron pipe substituted. In every respect it is believed that the design and workmanship employed in mounting and wiring the motors and control equipments under these steel cars is unequaled elsewhere in similar work up to the present time.

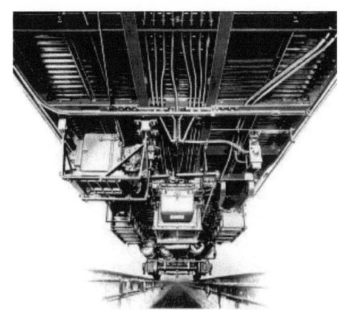

APPARATUS UNDER STEEL MOTOR CAR

The motors and car wiring are protected by a carefully planned system of fuses, the function of which is to melt and open the circuits, so cutting off power in case of failure of insulation.

Express trains and local trains alike are provided with a bus line, which interconnects the electrical supply to all cars and prevents interruption of the delivery of current to motors in case the collector shoes attached to any given car should momentarily fail to make contact with the third rail. At certain cross-overs this operates to prevent extinguishing the lamps in successive cars as the train passes from one track to another. The controller is so constructed that when the train is in motion the motorman is compelled to keep his hand upon it, otherwise the power is automatically cut off and the brakes are applied. This important safety device, which, in case a motorman be suddenly incapacitated at his post, will promptly stop the train, is a recent invention and is first introduced in practical service upon trains of the Interborough Company.

Heating and Lighting

All cars are heated and lighted by electricity. The heaters are placed beneath the seats, and special precautions have been taken to insure uniform distribution of the heat. The wiring for heaters and lights has been practically safe-guarded to avoid, so far as possible, all risk of short-circuit or fire, the wire used for the heater circuits being carried upon porcelain insulators from all woodwork by large clearances, while the wiring for lights is carried in metallic conduit. All lamp sockets are specially designed to prevent possibility of fire and are separated from the woodwork of the car by air spaces and by asbestos.

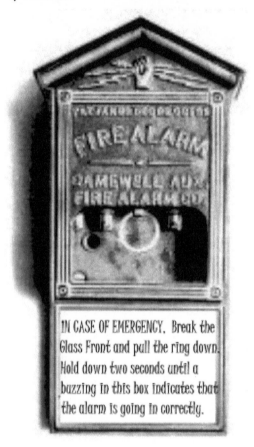

(FIRE ALARM)

The interior of each car is lighted by twenty-six 10-candle power lamps, in addition to four lamps provided for platforms and markers. The lamps for lighting the interior are carefully located, with a view to securing uniform and effective illumination.

CHAPTER VII

LIGHTING SYSTEM FOR PASSENGER STATIONS AND TUNNEL

In the initial preparation of plans, and more than a year before the accident which occurred in the subway system of Paris in August, 1903, the engineers of the Interborough Company realized the importance of maintaining lights in the subway independent of any temporary interruption of the power used for lighting the cars, and, in preparing their plans, they provided for lighting the subway throughout its length from a source independent of the main power supply. For this purpose three 1,250-kilowatt alternators direct-driven by steam turbines are installed in the power house, from which point a system of primary cables, transformers and secondary conductors convey current to the incandescent lamps used solely to light the subway. The alternators are of the three-phase type, making 1,200 revolutions per minute and delivering current at a frequency of 60 cycles per second at a potential of 11,000 volts. In the boiler plant and system of steam piping installed in connection with these turbine-driven units, provision is made for separation of the steam supply from the general supply for the 5,000 kilowatt units and for furnishing the steam for the turbine units through either of two alternative lines of pipe.

The 11,000 volt primary current is conveyed through paper insulated lead-sheathed cables to transformers, located in fireproof compartments adjacent to the platforms of the passenger stations. These transformers deliver current to two separate systems of secondary wiring, one of which is supplied at a potential of 120 volts and the other at 600 volts.

The general lighting of the passenger station platforms is effected by incandescent lamps supplied from the 120-volt secondary wiring circuits, while the lighting of the subway sections between adjacent stations is accomplished by incandescent lamps connected in series groups of five each and connected to the 600-volt lighting circuits. Recognizing the fact that in view of the precautions taken it is probable that interruptions of the alternating current lighting service will be infrequent, the possibility of such interruption is nevertheless provided for by installing upon the stairways leading to passenger station platforms, at the ticket booths and over the tracks in front of the platforms, a number of lamps which are connected to the contact rail circuit. This will provide light sufficient to enable passengers to see stairways and the edges of the station platforms in case of temporary failure of the general lighting system.

The general illumination of the passenger stations is effected by means of 32 c. p. incandescent lamps, placed in recessed domes in the ceiling. These are reinforced by 14 c. p. and 32 c. p. lamps, carried by brackets of ornate design where the construction of the station does not conveniently permit the use of ceiling lights. The lamps are enclosed in sand-blasted glass globes, and excellent distribution is secured by the use of reflectors.

The illustration on page 122 is produced from a photograph of the interior of one of the transformer cupboards and shows the transformer in place with the end bell of the high potential cable and the primary switchboard containing switches and enclosed fuses. The illustration on page 123 shows one of the secondary distributing switchboards which are located immediately behind the ticket booths, where they are under the control of the ticket seller.

TRANSFORMER COMPARTMENT IN PASSENGER STATION

In lighting the subway between passenger stations, it is desirable, on the one hand, to provide sufficient light for track inspection and to permit employees passing along the subway to see their way clearly and avoid obstructions; but, on the other hand, the lighting must not be so brilliant as to interfere with easy sight and recognition of the red, yellow, and green signal lamps of the block signal system. It is necessary also that the lights for general illumination be so placed that their rays shall not fall directly upon the eyes of approaching motormen at the head of trains nor annoy passengers who may be reading their papers inside the cars. The conditions imposed by these considerations are met in the four-track sections of the subway by placing a row of incandescent lamps between the north-bound local and express tracks and a similar row between the southbound local and express tracks. The lamps are carried upon brackets supported upon the iron columns of the subway structure, successive lamps in each row being 60 feet apart. They are located a few inches above the tops of the car windows and with reference to the direction of approaching trains the lamps in each row are carried upon the far side of the iron columns, by which expedient the eyes of the approaching motormen are sufficiently protected against their direct rays.

Lighting of the Power House

For the general illumination of the engine room, clusters of Nernst lamps are supported from the roof trusses and a row of single lamps of the same type is carried on the lower gallery about 25 feet from the floor. This is the first power house in America to be illuminated by these lamps. The quality of the light is unsurpassed and the general effect of the illumination most satisfactory and agreeable to the eye. In addition to the Nernst lamps, 16 c. p. incandescent lamps are placed upon the engines and along the galleries in places not conveniently reached by the general illumination. The basement also is lighted by incandescent lamps.

SECONDARY DISTRIBUTING SWITCHBOARD AT PASSENGER STATION

For the boiler room, a row of Nernst lamps in front of the batteries of boilers is provided, and, in addition to these, incandescent lamps are used in the passageways around the boilers, at gauges and at water columns. The basement of the boiler room, the pump room, the economizer floor, coal bunkers, and coal conveyers are lighted by incandescent lamps, while arc lamps are used around the coal tower and dock. The lights on the engines and those at gauge glasses and water columns and at the pumps are supplied by direct current from the 250-volt circuits. All other incandescent lamps and the Nernst lamps are supplied through transformers from the 60-cycle lighting system.

Emergency Signal System and Provision for Cutting Off Power from Contact Rail

In the booth of each ticket seller and at every manhole along the west side of the subway and its branches is placed a glass-covered box of the kind generally used in large American cities for fire alarm purposes. In case of accident in the subway which may render it desirable to cut off power from the contact rails, this result can be accomplished by breaking the glass front of the emergency box and pulling the hook provided. Special emergency circuits are so arranged that pulling the hook will instantly open all the circuit-breakers at adjacent sub-stations through which the contact rails in the section affected receive their supply of power. It will also instantly report the location of the trouble, annunciator gongs being located in the sub-stations from which power is supplied to the section, in the train dispatchers' offices and in the office of the General Superintendent, instantly intimating the number of the box which has been pulled. Automatic recording devices in train dispatchers' offices and in the office of the General Superintendent also note the number of the box pulled.

The photograph on page 120 shows a typical fire alarm box.

CHAPTER VIII

ROLLING STOCK—CARS, TRUCKS, ETC.

The determination of the builders of the road to improve upon the best devices known in electrical railroading and to provide an equipment unequaled on any interurban line is nowhere better illustrated than in the careful study given to the types of cars and trucks used on other lines before a selection was made of those to be employed on the subway.

All of the existing rapid transit railways in this country, and many of those abroad, were visited and the different patterns of cars in use were considered in this investigation, which included a study of the relative advantages of long and short cars, single and multiple side entrance cars and end entrance cars, and all of the other varieties which have been adopted for rapid transit service abroad and at home.

The service requirement of the New York subway introduces a number of unprecedented conditions, and required a complete redesign of all the existing models. The general considerations to be met included the following:

High schedule speeds with frequent stops.

Maximum carrying capacity for the subway, especially at times of rush hours, morning and evening.

Maximum strength combined with smallest permissible weight.

Adoption of all precautions calculated to reduce possibility of damage from either the electric circuit or from collisions.

The clearance and length of the local station platforms limited the length of trains, and tunnel clearances the length and width and height of the cars.

The speeds called for by the contract with the city introduced motive power requirements which were unprecedented in any existing railway service, either steam or electric, and demanded a minimum weight consistent with safety. As an example, it may be stated that an express train of eight cars in the subway to conform to the schedule speed adopted will require a nominal power of motors on the train of 2,000 horse power, with an average accelerating current at 600 volts in starting from a station stop of 325 amperes. This rate of energy absorption which corresponds to 2,500 horse power is not far from double that taken by the heaviest trains on trunk line railroads when starting from stations at the maximum rate of acceleration possible with the most powerful modern steam locomotives.

Such exacting schedule conditions as those mentioned necessitated the design of cars, trucks, etc., of equivalent strength to that found in steam railroad car and locomotive construction, so that while it was essential to keep down the weight of the train and individual cars to a minimum, owing to the frequent stops, it was equally as essential to provide the strongest and most substantial type of car construction throughout.

Owing to these two essentials which were embodied in their construction it can safely be asserted that the cars used in the subway represent the acme of car building art as it exists to-day, and that all available appliances for securing strength and durability in the cars and immunity from accidents have been introduced.

After having ascertained the general type of cars which would be best adapted to the subway service, and before placing the order for car equipments, it was decided to build sample cars embodying the approved principles of design. From these the management believed that the details of construction could be more perfectly determined than in any other way. Consequently, in the early part of 1902, two sample cars were built and equipped with a variety of appliances and furnishings so that the final type could be intelligently selected. From the tests conducted on these cars the adopted type of car which is described in detail below was evolved.

After the design had been worked out a great deal of difficulty was encountered in securing satisfactory contracts for proper deliveries, on account of the congested condition of the car building works in the country. Contracts were finally closed, however, in December, 1902, for 500 cars, and orders were distributed between four car-building firms. Of these cars, some 200, as fast as delivered, were placed in operation on the Second Avenue line of the Elevated Railway, in order that they might be thoroughly tested during the winter of 1903-4.

END VIEW OF STEEL PASSENGER CAR

In view of the peculiar traffic conditions existing in New York City and the restricted siding and yard room available in the subway, it was decided that one standard type of car for all classes of service would introduce the most flexible operating conditions, and for this reason would best suit the public demands at different seasons of the year and hours of the day. In order further to provide cars, each of which would be as safe as the others, it was essential that there should be no difference in constructional strength between the motor cars and the trail cars. All cars were therefore made of one type and can be used interchangeably for either motor or trail-car service.

The motor cars carry both motors on the same truck; that is, they have a motor truck at one end carrying two motors, one geared to each axle; the truck at the other end of the car is a "trailer" and carries no motive power.

SIDE VIEW OF STEEL PASSENGER CAR

Some leading distinctive features of the cars may be enumerated as follows:

(1.) The length is 51 feet and provides seating capacity for 52 passengers. This length is about 4 feet more than those of the existing Manhattan Elevated Railroad cars.

(2.) The enclosed vestibule platforms with sliding doors instead of the usual gates. The enclosed platforms will contribute greatly to the comfort and safety of passengers under subway conditions.

(3.) The anti-telescoping car bulkheads and platform posts. This construction is similar to that in use on Pullman cars, and has been demonstrated in steam railroad service to be an important safety appliance.

(4.) The steel underframing of the car, which provides a rigid and durable bed structure for transmitting the heavy motive power stresses.

(5.) The numerous protective devices against defects in the electrical apparatus.

(6.) Window arrangement, permitting circulation without draughts.

(7.) Emergency brake valve on truck operated by track trip.

(8.) Emergency brake valve in connection with master-controller.

The table on page 133 shows the main dimensions of the car, and also the corresponding dimensions of the standard car in use on the Manhattan Elevated Railway.

The general arrangement of the floor framing is well shown in the photograph on page 132. The side sills are of 6-inch channels, which are reinforced inside and out by white oak timbers. The center sills are 5-inch I-beams, faced on both sides with Southern pine. The end sills are also of steel shapes, securely attached to the side sills by steel castings and forgings.

The car body end-sill channel is faced with a white-oak filler, mortised to receive the car body end-posts and braced at each end by gusset plates. The body bolster is made up of two rolled steel plates bolted together at their ends and supported by a steel draw casting, the ends of which form a support for the center sills. The cross-bridging and needle-beams of 5-inch I-beams are unusually substantial. The flooring inside the car is double and of maple, with asbestos fire-felt between the layers, and is protected below by steel plates and "transite" (asbestos board).

The side framing of the car is of white ash, doubly braced and heavily trussed. There are seven composite wrought-iron carlines forged in shape for the roof, each sandwiched between two white ash carlines, and with white ash intermediate carlines. The platform posts are of compound construction with anti-telescoping posts of steel bar sandwiched between white ash posts at corners and centers of vestibuled platforms. These posts are securely bolted to the steel longitudinal sills, the steel anti-telescoping plate below the floor, and to the hood of the bow which serves to reinforce it. This bow is a heavy steel angle in one piece, reaching from plate to plate and extending back into the car 6 feet on each side. By this construction it is believed that the car framing is practically indestructible. In case of accident, if one platform should ride over another, eight square inches of metal would have to be sheared off the posts before the main body of the car would be reached, which would afford an effective means of protection.

EXTERIOR VIEW—STEEL CAR FRAMING

The floor is completely covered on the underside with 1/4-inch asbestos transite board, while all parts of the car framing, flooring, and sheathing are covered with fire-proofing compound. In addition, all spaces above the motor truck in the floor framing, between sills and bridging, are protected by plates of No. 8 steel and 1/4-inch roll fire-felt extending from the platform end sill to the bolster.

Car Wiring

The precautions to secure safety from fire consists generally in the perfected arrangement and installation of the electrical apparatus and the wiring. For the lighting circuits a flexible steel conduit is used, and a special junction box. On the side and upper roofs, over these conduits for the lighting circuits, a strip of sheet iron is securely nailed to the roof boards before the canvas is applied. The wires under the floor are carried in ducts moulded into suitable forms of asbestos compound. Special precautions have been taken with the insulation of the wires, the specifications calling for, first, a layer of paper, next, a layer of rubber, and then a layer of cotton saturated with a weather-proof compound, and outside of this a layer of asbestos. The hangers supporting the rheostats under the car body are insulated with wooden blocks, treated by a special process, being dried out in an oven and then soaked in an insulating compound, and covered with 1/4-inch "transite" board. The rheostat boxes themselves are also insulated from the angle iron supporting them. Where the wires pass through the flooring they are hermetically sealed to prevent the admission of dust and dirt.

At the forward end of what is known as the No. 1 end of the car all the wires are carried to a slate switchboard in the motorman's cab. This board is 44 x 27 inches, and is mounted directly back of the motorman. The window space occupied by this board is ceiled up and the space back of the panels is boxed in and provided with a door of steel plate, forming a box, the cover, top, bottom, and sides of which are lined with electrobestos 1/2-inch thick. All of the switches and fuses, except the main trolley fuse and bus-line fuse, which are encased and placed under the car, are carried on this switchboard. Where the wires are carried through the floor or any partition, a steel chute, lined with electrobestos, is used to protect the wires against mechanical injury. It will be noted from the above that no power wiring, switches, or fuses are placed in the car itself, all such devices being outside in a special steel insulated compartment.

A novel feature in the construction of these cars is the motorman's compartment and vestibule, which differs essentially from that used heretofore, and the patents are owned by the Interborough Company. The cab is located on the platform, so that no space within the car is required; at the same time the entire platform space is available for ingress and egress except that on the front platform of the first car, on which the passengers would not be allowed in any case. The side of the cab is formed by a door which can be placed in three positions. When in its mid-position it encloses a part of the platform, so as to furnish a cab for the motorman, but when swung parallel to the end sills it encloses the end of the platform, and this would be its position on the rear platform of the rear car. The third position is when it is swung around to an arc of 180 degrees, when it can be locked in position against the corner vestibule post enclosing the master controller. This would be its position on all platforms except on the front of the front car or the rear of the rear car of the train.

The platforms themselves are not equipped with side gates, but with doors arranged to slide into pockets in the side framing, thereby giving up the entire platform to the passengers. These doors are closed by an overhead lever system. The sliding door on the front platform of the first car may be partly opened and secured in this position by a bar, and thus serve as an arm-rest for the motorman. The doors close against an air-cushion stop, making it impossible to clutch the clothing or limbs of passengers in closing.

INTERIOR VIEW—SKELETON FRAMING OF STEEL CAR

Pantagraph safety gates for coupling between cars are provided. They are constructed so as to adjust themselves to suit the various positions of adjoining cars while passing in, around, and out of curves of 90 feet radius.

On the door leading from the vestibule to the body of the car is a curtain that can be automatically raised and lowered as the door is opened or closed to shut the light away from the motorman. Another attachment is the peculiar handle on the sliding door. This door is made to latch so that it cannot slide open with the swaying of the car, but the handle is so constructed that when pressure is applied upon it to open the door, the same movement will unlatch it.

Entering the car, the observer is at once impressed by the amount of room available for passengers. The seating arrangements are similar to the elevated cars, but the subway coaches are longer and wider than the Manhattan, and there are two additional seats on each end. The seats are all finished in rattan. Stationary crosswise seats are provided after the Manhattan pattern, at the center of the car. The longitudinal seats are 17-3/4 inches deep. The space between the longitudinal seats is 4 feet 5 inches.

The windows have two sashes, the lower one being stationary, while the upper one is a drop sash. This arrangement reverses the ordinary practice, and is desirable in subway operation and to insure safety and comfort to the passengers. The side windows in the body of the car, also the end windows and end doors, are provided with roll shades with pinch-handle fixtures.

INTERIOR VIEW OF PROTECTED WOODEN CAR

The floors are covered with hard maple strips, securely fastened to the floor with ovalhead brass screws, thus providing a clean, dry floor for all conditions of weather.

Six single incandescent lamps are placed on the upper deck ceiling, and a row of ten on each side deck ceiling is provided. There are two lamps placed in a white porcelain dome over each platform, and the pressure gauge is also provided with a miniature lamp.

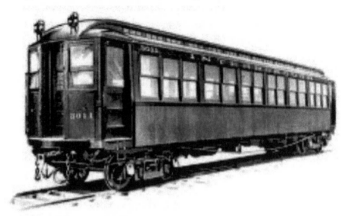

EXTERIOR VIEW—PROTECTED WOODEN CAR, SHOWING COPPER SIDES

The head linings are of composite board. The interior finish is of mahogany of light color. A mahogany handrail extends the full length of the clerestory on each side of the car, supported in brass sockets at the ends and by heavy brass brackets on each side. The handrail on each side of the car carries thirty-eight leather straps.

Each ventilator sash is secured on the inside to a brass operating arm, manipulated by means of rods running along each side of the clerestory, and each rod is operated by means of a brass lever, having a fulcrum secured to the inside of the clerestory.

All hardware is of bronze, of best quality and heavy pattern, including locks, pulls, handles, sash fittings, window guards, railing brackets and sockets, bell cord thimbles, chafing strips, hinges, and all other trimmings. The upright panels between the windows and the corner of the car are of plain mahogany, as are also the single post pilasters, all of which are decorated with marquetry inlaid. The end finish is of mahogany, forming a casing for the end door.

FRAMING OF PROTECTED WOODEN CAR

Steel Cars

At the time of placing the first contract for the rolling stock of the subway, the question of using an all-steel car was carefully considered by the management. Such a type of car, in many respects, presented desirable features for subway work as representing the ultimate of absolute incombustibility. Certain practical reasons, however, prevented the adoption of an all-steel car in the spring of 1902 when it became necessary to place the orders mentioned above for the first 500 cars. Principal among these reasons was the fact that no cars of this kind had ever been constructed, and as the car building works of the country were in a very congested condition all of the larger companies declined to consider any standard specifications even for a short-time delivery, while for cars involving the extensive use of metal the question was impossible of immediate solution. Again, there were a number of very serious mechanical difficulties to be studied and overcome in the construction of such a car, such as avoidance of excessive weight, a serious element in a rapid transit service, insulation from the extremes of heat and cold, and the prevention of undue noise in operation. It was decided, therefore, to bend all energies to the production of a wooden car with sufficient metal for strength and protection from accident, i. e., a stronger, safer, and better constructed car than had heretofore been put in use on any electric railway in the world. These properties it is believed are embodied in the car which has just been described.

METAL UNDERFRAME OF PROTECTED WOODEN CAR

The plan of an all-metal car, however, was not abandoned, and although none was in use in passenger service anywhere, steps were immediately taken to design a car of this type and conduct the necessary tests to determine whether it would be suitable for railway service. None of the car-building companies was willing to undertake the work, but the courteous coöperation of the Pennsylvania Railroad Company was secured in placing its manufacturing facilities at Altoona at the disposal of the Interborough Rapid Transit Railway Company. Plans were prepared for an all-metal car, and after about fourteen months of work a sample type was completed in December, 1903, which was in every way creditable as a first attempt.

The sample car naturally embodied some faults which only experience could correct, the principal one being that the car was not only too heavy for use on the elevated lines of the company, but attained an undesirable weight for subway operation. From this original design, however, a second design involving very original features has been worked out, and a contract has been given by the Interborough Company for 200 all-steel cars, which are now being constructed. While the expense of producing this new type of car has obviously been great, this consideration has not influenced the management of the company in developing an equipment which promised the maximum of operating safety.

END VIEW OF MOTOR TRUCK

The General Arrangements

The general dimensions of the all-steel car differ only slightly from those of the wooden car. The following table gives the dimensions of the two cars, and also that of the Manhattan Railway cars:

	Wooden Cars.	All-Steel Cars.	Manhattan Cars.
Length over body corner posts,	42 ft. 7 ins.	41 ft. 1/2 in.	39 ft. 10 ins.
Length over buffers,	51 ft. 2 ins.	51 ft. 2 ins.	47 ft. 1 in.
Length over draw-bars,	51 ft. 5 ins.	51 ft. 5 ins.	47 ft. 4 ins.
Width over side sills,	8 ft. 8-3/8 ins.	8 ft. 6-3/4 ins.	8 ft. 6 ins.

Width over sheathing,	8 ft. 10 ins.	8 ft. 7 ins.	8 ft. 7 ins.
Width over window sills,	8 ft. 11-7/8 ins.	9 ft. 1/2 in.	8 ft. 9 ins.
Width over battens,	8 ft. 10-3/4 ins.	8 ft. 7-1/4 ins.	8 ft. 7-7/8 ins.
Width over eaves,	8 ft. 8 ins.	8 ft. 8 ins.	8 ft. 9-1/2 ins.
Height from under side of sill to top of plate,	7 ft. 3-1/8 ins.	7 ft. 1 in.	7 ft. 3 ins.
Height of body from under side of center sill to top of roof,	8 ft. 9-7/8 ins.	8 ft. 9-7/8 ins.	9 ft. 5-7/8 ins.
Height of truck from rail to top of truck center plate (car light),	2 ft. 8 ins.	2 ft. 8 ins.	2 ft. 5-3/4 ins.
Height from top of rail to underside of side sill at truck center (car light),	3 ft. 1-1/8 ins.	3 ft. 2-1/8 ins.	3 ft. 3-1/4 ins.
Height from top of rail to top of roof not to exceed (car light),	12 ft. 3/4 in.	12 ft. 0 in.	12 ft. 10-1/2 ins.

The general frame plan of the all-steel car is clearly shown by the photograph on page 128. As will be seen, the floor framing is made up of two center longitudinal 6-inch I-beams and two longitudinal 5 x 3-inch steel side angles, extending in one piece from platform-end sill to platform-end sill. The end sills are angles and are secured to the side and center sills by cast-steel brackets, and in addition by steel anti-telescoping plates, which are placed on the under side of the sills and riveted thereto. The flooring is of galvanized, corrugated sheet iron, laid across the longitudinal sills and secured to longitudinal angles by rivets. This corrugated sheet holds the fireproof cement flooring called "monolith." On top of this latter are attached longitudinal floor strips for a wearing surface. The platform flooring is of steel plate covered with rubber matting cemented to the same. The side and end frame is composed of single and compound posts made of steel angles or T's and the roof framing of wrought-iron carlines and purlines. The sides of the cars are double and composed of steel plates on the outside, riveted to the side posts and belt rails, and lined with electrobestos. The outside roof is of fireproof composite board, covered with canvas. The headlinings are of fireproof composite, faced with aluminum sheets. The mouldings throughout are of aluminum. The wainscoting is of "transite" board and aluminum, and the end finish and window panels are of aluminum, lined with asbestos felt. The seat frames are of steel throughout, as are also the cushion frames. The sash is double, the lower part being stationary and the upper part movable. The doors are of mahogany, and are of the sliding type and are operated by the door operating device already described.

SIDE VIEW OF MOTOR TRUCK

Trucks

Two types of trucks are being built, one for the motor end, the other for the trailer end of the car. The following are the principal dimensions of the trucks:

	Motor Truck.	**Trailer Truck.**
Gauge of track,	4 ft. 8-1/2 ins.	4 ft. 8-1/2 ins.
Distance between backs of wheel flanges,	4 ft. 5-3/8 ins.	4 ft. 5-3/8 ins.
Height of truck center plate above rail, car body loaded with 15,000 pounds,	30 ins.	30 ins.

Height of truck side bearings above rail, car body loaded,	34 ins.	34 ins.
Wheel base of truck,	6 ft. 8 ins.	5 ft. 6 ins.
Weight on center plate with car body loaded, about	27,000 lbs.	
Side frames, wrought-iron forged,	2-1/2 ins. x 4 ins.	1-1/2 ins. x 3 ins.
Pedestals, wrought-iron forged,		
Center transom, steel channel,		
Truck bolster,	cast steel.	wood and iron.
Equalizing bars, wrought iron,		
Center plate, cast steel,		
Spring plank, wrought iron,	1 in. x 3 ins.	white oak.
Bolster springs, elliptic, length,	30 ins.	32 ins.
Equalizing springs, double coil, outside dimensions,	4-7/8 ins. x 7-1/2 ins.	3-5/8 ins. x 6 ins.
Wheels, cast steel spoke center, steel tired, diameter,	33-3/4 ins.	30 ins.
Tires, tread M. C. B. Standard,	2-5/8 ins. x 5-1/4 ins.	2-5/8 ins. x 5-1/4 ins.
Axles, diameter at center,	6-1/2 ins.	4-3/4 ins.
Axles, diameter at gear seat,	7-13/16 ins.	
Axles, diameter at wheel seat,	7-3/4 ins.	5-3/4 ins.
Journals,	5 ins. x 9 ins.	4-1/4 ins. x 8

		ins.
Journal boxes, malleable iron, M. C. B. Standard,		

Both the motor and the trailer trucks have been designed with the greatest care for severe service, and their details are the outcome of years of practical experience.

CHAPTER IX

SIGNAL SYSTEM

Early in the development of the plans for the subway system in New York City, it was foreseen that the efficiency of operation of a road with so heavy a traffic as is being provided for would depend largely upon the completeness of the block signaling and interlocking systems adopted for spacing and directing trains. On account of the importance of this consideration, not only for safety of passengers, but also for conducting operation under exacting schedules, it was decided to install the most complete and effective signaling system procurable. The problem involved the prime consideration of:

Safety and reliability.

Greatest capacity of the lines consistent with the above.

Facility of operation under necessarily restricted yard and track conditions.

In order to obtain the above desiderata it was decided to install a complete automatic block signal system for the high-speed routes, block protection for all obscure points on the low-speed routes, and to operate all switches both for line movements and in yards by power from central points. This necessarily involved the interconnection of the block and switch movements at many locations and made the adoption of the most flexible and compact appliances essential.

Of the various signal systems in use it was found that the one promising entirely satisfactory results was the electro-pneumatic block and interlocking system, by which power in any quantity could be readily conducted in small pipes any distance and utilized in compact apparatus in the most restricted spaces. The movements could be made with the greatest promptness and certainty and interconnected for the most complicated situations for safety. Moreover, all essential details of the system had been worked out in years of practical operation on important trunk lines of railway, so that its reliability and efficiency were beyond question.

The application of such a system to the New York subway involved an elaboration of detail not before attempted upon a railway line of similar length, and the contract for its installation is believed to be the largest single order ever given to a signal manufacturing company.

In the application of an automatic block system to an electric railway where the rails are used for the return circuit of the propulsion current, it is necessary to modify the system as usually applied to a steam railway and introduce a track circuit control that will not be injuriously influenced by the propulsion current. This had been successfully accomplished for moderately heavy electric railway traffic in the Boston elevated installation, which was the first electric railway to adopt a complete automatic block signal system with track circuit control.

The New York subway operation, however, contemplated traffic of unprecedented density and consequent magnitude of the electric currents employed, and experience with existing track circuit control systems led to the conclusion that some modification in apparatus was essential to prevent occasional traffic delays.

The proposed operation contemplates a possible maximum of two tracks loaded with local trains at one minute intervals, and two tracks with eight car express trains at two minute intervals, the latter class of trains requiring at times as much as 2,000 horse power for each train in motion. It is readily seen, then, that combinations of trains in motion may at certain times occur which will throw enormous demands for power upon a given section of the road. The electricity conveying this power flows back through the track rails to the power station and in so doing is subject to a "drop" or loss in the rails which varies in amount according to the power demands. This causes disturbances in the signal-track circuit in proportion to the amount of "drop," and it was believed that under the extreme condition above mentioned the ordinary form of track circuit might prove unreliable and cause delay to traffic. A solution of the difficulty was suggested, consisting in the employment of a current in the signal track circuit which would have such characteristic differences from that used to propel the trains as would operate selectively upon an apparatus which would in turn control the signal. Alternating current supplied this want on account of its inductive properties, and was adopted, after a demonstration of its practicability under similar conditions elsewhere.

**FRONT VIEW OF BLOCK SIGNAL POST, SHOWING
LIGHTS, INDICATORS AND TRACK STOP**

After a decision was reached as to the system to be employed, the arrangement of the block sections was considered from the standpoint of maximum safety and maximum traffic capacity, as it was realized that the rapidly increasing traffic of Greater New York would almost at once tax the capacity of the line to its utmost.

The usual method of installing automatic block signals in the United States is to provide home and distant signals with the block sections extending from home signal to home signal; that is, the block sections end at the home signals and do not overlap each other. This is also the arrangement of block sections where the telegraph block or controlled manual systems are in use. The English block systems, however, all employ overlaps. Without the overlap, a train in passing from one block section to the other will clear the home signals for the section in the rear, as soon as the rear of the train has passed the home signal of the block in which it is moving. It is thus possible for a train to stop within the block and within a few feet of this home signal. If, then, a following train should for any reason overrun this home signal, a collision would result. With the overlap system, however, a train may stop at any point in a block section and still have the home signal at a safe stopping distance in the rear of the train.

Conservative signaling is all in favor of the overlap, on account of the safety factor, in case the signal is accidentally overrun. Another consideration was the use of automatic train stops. These stops are placed at the home signals, and it is thus essential that a stopping distance should be afforded in advance of the home signal to provide for stopping the train to which the brake had been applied by the automatic stop.

Ordinarily, the arrangement of overlap sections increases the length of block sections by the length of the overlap, and as the length of the section fixed the minimum spacing of trains, it was imperative to make the blocks as short as consistent with safety, in order not to cut down the carrying capacity of the railway. This led to a study of the special problem presented by subway signaling and a development of a blocking system upon lines which it is believed are distinctly in advance of anything heretofore done in this direction.

**REAR VIEW OF BLOCK SIGNAL POST, SHOWING
TRANSFORMER AND INSTRUMENT CASES WITH DOORS
OPEN**

Block section lengths are governed by speed and interval between trains. Overlap lengths are determined by the distance in which a train can be stopped at a maximum speed. Usually the block section length is the distance between signals, plus the overlap; but where maximum traffic capacity is desired the block section length can be reduced to the length of two overlaps, and this was the system adopted for the Interborough. The three systems of blocking trains, with and without overlaps, is shown diagramatically on page 143, where two successive trains are shown at the minimum distances apart for "clear" running for an assumed stopping distance of 800 feet. The system adopted for the subway is shown in line "C," giving the least headway of the three methods.

PNEUMATIC TRACK STOP, SHOWING STOP TRIGGER IN UPRIGHT POSITION

The length of the overlap was given very careful consideration by the Interborough Rapid Transit Company, who instituted a series of tests of braking power of trains; from these and others made by the Pennsylvania Railroad Company, curves were computed so as to determine the distance in which trains could be stopped at various rates of speed on a level track, with corrections for rising and falling to grades up to 2 per cent. Speed curves were then plotted for the trains on the entire line, showing at each point the maximum possible speed, with the gear ratio of the motors adopted. A joint consideration of the speeds, braking efforts, and profile of the road were then used to determine at each and every point on the line the minimum allowable distance between trains, so that the train in the rear could be stopped by the automatic application of the brakes before reaching a train which might be standing at a signal in advance; in other words, the length of the overlap section was determined by the local conditions at each point.

In order to provide for adverse conditions the actual braking distances was increased by 50 per cent.; for example, the braking distance of a train moving 35 miles an hour is 465 feet, this would be increased 50 per cent. and the overlap made not less than 697 feet. With this length of overlap the home signals could be located 697 feet apart, and the block section length would be double this or 1394 feet. The average length of overlaps, as laid out, is about 800 feet, and the length of block sections double this, or 1,600 feet.

\VIEW UNDER CAR, SHOWING TRIGGER ON TRUCK IN POSITION TO ENGAGE WITH TRACK STOP

The protection provided by this unique arrangement of signals is illustrated on page 143. Three positions of train are shown:

"A." MINIMUM distance between trains: The first train has just passed the home signal, the second train is stopped by the home signal in the rear; if this train had failed to stop at this point, the automatic stop would have applied the air brake and the train would have had the overlap distance in which to stop before it could reach the rear of the train in advance; therefore, under the worst conditions, no train can get closer to the train in advance than the length of the overlap, and this is always a safe stopping distance.

"B." CAUTION distance between train: The first train in same position as in "A," the second train at the third home signal in the rear; this signal can be passed under caution, and this distance between trains is the caution distance, and is always equal to the length of the block section, or two overlaps.

"C." CLEAR distance between trains: First train in same position as in "A," second train at the fourth home signal in the rear; at this point both the home and distant signals are clear, and the distance between the trains is now the clear running distance; that is, when the trains are one block section plus an overlap apart they can move under clear signal, and this distance is used in determining the running schedule. It will be noted in "C" that the first train has the following protection: Home signals 1 and 2 in stop position, together with the automatic stop at signal 2 in position to stop a train, distant signal 1, 2, and 3 all at caution, or, in other words, a train that has stopped is always protected by two home signals in its rear, and by three caution signals, in addition to this an automatic stop placed at a safe stopping distance in the rear of the train.

ELECTRO-PNEUMATIC INTERLOCKING MACHINE ON STATION PLATFORM

SPECIAL INTERLOCKING SIGNAL CABIN SOUTH OF BROOKLYN BRIDGE STATION

Description of Block Signaling System

The block signaling system as installed consists of automatic overlapping system above described applied to the two express tracks between City Hall and 96th Street, a distance of six and one-half miles, or thirteen miles of track; and to the third track between 96th and 145th Streets on the West Side branch, a distance of two and one-half miles. This third track is placed between the two local tracks, and will be used for express traffic in both directions, trains moving toward the City Hall in the morning and in the opposite direction at night; also the two tracks from 145th Street to Dyckman Street, a distance of two and one-half miles, or five miles of track. The total length of track protected by signals is twenty-four and one-half miles.

The small amount of available space in the subway made it necessary to design a special form of the signal itself. Clearances would not permit of a "position" signal indication, and, further, a position signal purely was not suitable for the lighting conditions of the subway. A color signal was therefore adopted conforming to the adopted rules of the American Railway Association. It consists of an iron case fitted with two white lenses, the upper being the home signal and the lower the distant. Suitable colored glasses are mounted in slides which are operated by pneumatic cylinders placed in the base of the case. Home and dwarf signals show a red light for the danger or "stop" indication. Distant signals show a yellow light for the "caution" indication. All signals show a green light for the "proceed" or clear position. Signals in the subway are constantly lighted by two electric lights placed back of each white lens, so that the lighting will be at all times reliable.

On the elevated structure, semaphore signals of the usual type are used. The signal lighting is supplied by a special alternating current circuit independent of the power and general lighting circuits.

A train stop or automatic stop of the Kinsman system is used at all block signals, and at many interlocking signals. This is a device for automatically applying the air brakes to the train if it should pass a signal in the stop position. This is an additional safeguard only to be brought into action when the danger indication has for any reason been disregarded, and insures the maintenance of the minimum distance between trains as provided by the overlaps established.

Great care has been given to the design, construction, and installation of the signal apparatus, so as to insure reliability of operation under the most adverse conditions, and to provide for accessibility to all the parts for convenience in maintenance. The system for furnishing power to operate and control the signals consists of the following:

Two 500-volt alternating current feed mains run the entire length of the signal system. These mains are fed by seven direct-current motor-driven generators operated in multiple located in the various sub-power stations. Any four of these machines are sufficient to supply the necessary current for operating the system. Across these alternating mains are connected the primary coils of track transformers located at each signal, the secondaries of which supply current of about 10 volts to the rails of the track sections. Across the rails at the opposite end of the section is connected the track relay, the moving element of which operates a contact. This contact controls a local direct-current circuit operating, by compressed air, the signal and automatic train stop.

Direct current is furnished by two mains extending the length of the system, which are fed by eight sets of 16-volt storage batteries in duplicate. These batteries are located in the subway at the various interlocking towers, and are charged by motor generators, one of which is placed at each set of batteries. These motor generators are driven by direct current from the third rail and deliver direct current of 25 volts.

The compressed air is supplied by six air compressors, one located at each of the following sub-stations: Nos. 11, 12, 13, 14, 16, and 17. Three of these are reserve compressors. They are motor-driven by direct-current motors, taking current from the direct-current buss bars at sub-stations at from 400 to 700 volts. The capacity of each compressor is 230 cubic feet.

MAIN LINE, PIPING AND WIRING FOR BLOCK AND INTERLOCKING SYSTEM, SHOWING JUNCTION BOX ON COLUMN

The motor-driven air compressors are controlled by a governor which responds to a variation of air pressure of five pounds or less. When the pressure has reached a predetermined point the machine is stopped and the supply of cooling water shut off. When the pressure has fallen a given amount, the machine is started light, and when at full speed the load is thrown on and the cooling water circulation reëstablished. Oiling of cylinders and bearings is automatic, being supplied only while the machines are running.

Two novel safety devices having to do especially with the signaling may be here described. The first is an emergency train stop. It is designed to place in the hands of station attendants, or others, the emergency control of signals. The protection afforded is similar in principle to the emergency brake handle found in all passenger cars, but operates to warn all trains of an extraneous danger condition. It has been shown in electric railroading that an accident to apparatus, perhaps of slight moment, may cause an unreasoning panic, on account of which passengers may wander on adjoining tracks in face of approaching trains. To provide as perfectly as practicable for such conditions, it has been arranged to loop the control of signals into an emergency box set in a conspicuous position in each station platform. The pushing of a button on this box, similar to that of the fire-alarm signal, will set all signals immediately adjacent to stations in the face of trains approaching, so that all traffic may be stopped until the danger condition is removed.

The second safety appliance is the "section break" protection. This consists of a special emergency signal placed in advance of each separate section of the third rail; that is, at points where trains move from a section fed by one sub-station to that fed by another. Under such conditions the contact shoes of the train temporarily span the break in the third rail. In case of a serious overload or ground on one section, the train-wiring would momentarily act as a feeder for the section, and thus possibly blow the train fuses and cause delay. In order, therefore, to prevent trains passing into a dangerously overloaded section, an overload relay has been installed at each section break to set a "stop" signal in the face of an approaching train, which holds the train until the abnormal condition is removed.

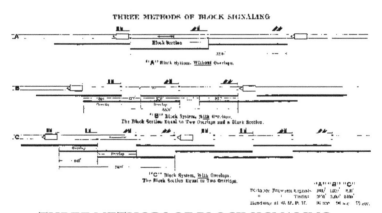

THREE METHODS OF BLOCK SIGNALING

- 175 -

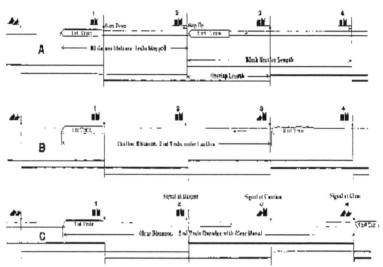

DIAGRAM OF OVERLAPPING BLOCK SIGNAL SYSTEM ILLUSTRATING POSSIBLE POSITIONS OF TRAINS RUNNING UNDER SAME

Interlocking System

The to-and-fro movement of a dense traffic on a four-track railway requires a large amount of switching, especially when each movement is complicated by junctions of two or more lines. Practically every problem of trunk line train movement, including two, three, and four-track operation, had to be provided for in the switching plants of the subway. Further, the problem was complicated by the restricted clearances and vision attendant upon tunnel construction. It was estimated that the utmost flexibility of operation should be provided for, and also that every movement be certain, quick, and safe.

All of the above, which are referred to in the briefest terms only, demanded that all switching movements should be made through the medium of power-operated interlocking plants. These plants in the subway portions of the line are in all cases electro-pneumatic, while in the elevated portions of the line mechanical interlocking has been, in some cases, provided.

A list of the separate plants installed will be interesting, and is given below:

Location.	Interlocking Machines.	Working Levers.
MAIN LINE.		
City Hall,	3	32
Spring Street,	2	10
14th Street,	2	16
18th Street,	1	4
42d Street,	2	15
72d Street	2	15
96th Street	2	19
WEST SIDE BRANCH.		
100th Street,	1	6
103d Street,	1	6
110th Street,	2	12
116th Street,	2	12
Manhattan Viaduct,	1	12
137th Street,	2	17
145th Street,	2	19
Dyckman Street,	1	12
216th Street,	1	14
EAST SIDE BRANCH.		
135th Street,	2	6
Lenox Junction,	1	7
145th Street,	1	9
Lenox Avenue Yard,	1	35
Third and Westchester Avenue Junction,	1	13

St. Anna Avenue,	1	24
Freeman Street,	1	12
176th Street,	2	66
	——	——
Total,	37	393

The total number of signals, both block and interlocking, is as follows:

Home signals,	354
Dwarf signals,	150
Distant signals,	187
	——
Total,	691
Total number of switches,	224

It will be noted that in the case of the City Hall Station three separate plants are required, all of considerable size, and intended for constant use for a multiplicity of movements. It is, perhaps, unnecessary to state that all the mechanism of these important interlocking plants is of the most substantial character and provided with all the necessary safety appliances and means for rapidly setting up the various combinations. The interlocking machines are housed in steel concrete "towers," so that the operators may be properly protected and isolated in the performance of their duties.

CHAPTER X

SUBWAY DRAINAGE

The employment of water-proofing to the exterior surfaces of the masonry shell of the tunnel, which is applied to the masonry, almost without a break along the entire subway construction, has made it unnecessary to provide an extensive system of drains, or sump pits, of any magnitude, for the collection and removal of water from the interior of the tunnel.

On the other hand, however, at each depression or point where water could collect from any cause, such as by leakage through a cable manhole cover or by the breaking of an adjacent water pipe, or the like, a sump pit or drain has been provided for carrying the water away from the interior of the tunnel.

For all locations, where such drains, or sump pits, are located above the line of the adjacent sewer, the carrying of the water away has been easy to accomplish by employing a drain pipe in connection with suitable traps and valves.

In other cases, however, where it is necessary to elevate the water, the problem has been of a different character. In such cases, where possible, at each depression where water is liable to collect, a well, or sump pit, has been constructed just outside the shell of the tunnel. The bottom of the well has been placed lower than the floor of the tunnel, so that the water can flow into the well through a drain connecting to the tunnel.

Each well is then provided with a pumping outfit; but in the case of these wells and in other locations where it is necessary to maintain pumping devices, it has not been possible to employ a uniform design of pumping equipment, as the various locations offer different conditions, each employing apparatus best suited to the requirements.

In no case, except two, is an electric pump employed, as the employment of compressed air was considered more reliable.

The several depressions at which it is necessary to maintain a pumping plant are enumerated as follows:

No. 1—Sump at the lowest point on City Hall Loop.

No. 2—Sump at intersection of Elm and White Streets.

No. 3—Sump at 38th Street in the Murray Hill Tunnel.

No. 4—Sump at intersection of 46th Street and Broadway.

No. 5—Sump at intersection of 116th Street and Lenox Avenue.

No. 6—Sump at intersection of 142d Street and Lenox Avenue.

No. 7—Sump at intersection of 147th Street and Lenox Avenue.

No. 8—Sump at about 144th Street in Harlem River approach.

No. 9—Sump at the center of the Harlem River Tunnel.

No. 10—Sump at intersection of Gerard Avenue and 149th Street.

In addition to the above mentioned sumps, where pumping plants are maintained, it is necessary to maintain pumping plants at the following points:

Location No. 1—At the cable tunnel constructed under the Subway at 23d Street and Fourth Avenue.

Location No. 2—At the sub-subway at 42d Street and Broadway.

Location No. 3—At the portal of the Lenox Avenue extension at 148th Street.

Location No. 4—At the southerly end of the Harlem River tube.

Location No. 5—At the northerly end of the Harlem River tube.

Location No. 6—At the portal at Bergen Avenue and 149th Street.

In the case of the No. 1 sump a direct-connected electric triple-plunger pump is employed, situated in a pump room about 40 feet distant from the sump pit. In the case of Nos. 2, 4, and 7 sumps, automatic air lifts are employed. This apparatus is placed in those sump wells which are not easily accessible, and the air lift was selected for the reason that no moving parts are conveyed in the air-lift construction other than the movable ball float and valve which control the device. The air lift consists of concentric piping extending several feet into the ground below the bottom of the well, and the water is elevated by the air producing a rising column of water of less specific weight than the descending column of water which is in the pipe extending below the bottom of the sump well.

In the case of Nos. 3 and 5 sumps, and for Location No. 1, automatic air-operated ejectors have been employed, for the reason that the conditions did not warrant the employment of air lifts or electric or air-operated pumps.

In the case of Nos. 6, 8, 9, and 10 sumps and for Locations Nos. 2, 4, and 5, air-operated reciprocating pumps will be employed. These pumps will be placed in readily accessible locations, where air lifts could not be used, and this type of pump was selected as being the most reliable device to employ.

In the case of Location No. 3, where provision has to be made to prevent a large amount of yard drainage, during a storm, from entering the tunnel where it descends from the portal, it was considered best to employ large submerged centrifugal pumps, operated by reciprocating air engines. Also for the portal, at Location No. 6, similar centrifugal pumps will be employed, but as compressed air is not available at this point, these pumps will be operated by electric motors.

The air supply to the air-operating pumping devices will be independent from the compressed air line which supplies air to the switch and signal system, but break-down connections will be made between the two systems, so that either system can help the other out in case of emergency.

A special air-compressor plant is located at the 148th Street repair shop, and another plant within the subway at 41st Street, for supplying air to the pumps, within the immediate locality of each compressor plant. For the more remote pumps, air will be supplied by smaller air compressors located within passenger stations. In one case, for the No. 2 sump, air will be taken from the switch and signal air-compressor plant located at the No. 11 sub-station.

CHAPTER XI

REPAIR AND INSPECTION SHED

While popularly and not inaccurately known as the "Subway System," the lines of the Interborough Company comprise also a large amount of trackage in the open air, and hence the rolling stock which has already been described is devised with the view to satisfying all the peculiar and special conditions thus involved. A necessary corollary is the requirement of adequate inspection and repair shops, so that all the rolling stock may at all times be in the highest state of efficiency; and in this respect the provision made by the company has been lavish and liberal to a degree.

The repair and inspection shop of the Interborough Rapid Transit Company adjoins the car yards of the company and occupies the entire block between Seventh Avenue on the west, Lenox Avenue and the Harlem River on the east, 148th Street on the south, and 149th Street on the north. The electric subway trains will enter the shops and car yard by means of the Lenox Avenue extension, which runs directly north from the junction at 142d Street and Lenox Avenue of the East Side main line. The branch leaves the main line at 142d Street, gradually approaches the surface, and emerges at about 147th Street.

General Arrangement

The inspection shed is at the southern end of the property and occupies an area of approximately 336 feet by 240 feet. It is divided into three bays, of which the north bay is equipped with four tracks running its entire length, and the middle bay with five tracks. The south bay contains the machine-tool equipment, and consists of eighteen electrically driven machines, locker and wash rooms, heating boilers, etc., and has only one track extending through it.

Construction

The construction of the inspection shops is that which is ordinarily known as "reinforced concrete," and no wood is employed in the walls or roof. The building is a steel structure made up of four rows of center columns, which consist of twenty-one bays of 16 feet each, supporting the roof trusses. The foundations for these center columns are concrete piers mounted on piles. After the erection of the steel skeleton, the sides of the building and the interior walls are constructed by the use of 3/4-inch furring channels, located 16 inches apart, on which are fastened a series of expanded metal laths. The concrete is then applied to these laths in six coats, three on each side, and termed respectively the scratch coat, the rough coat, and the fining coat. In the later, the concrete is made with white sand, to give a finished appearance to the building.

The roof is composed of concrete slabs, reinforced with expanded metal laths and finished with cement and mortar. It is then water-proofed with vulcanite water-proofing and gravel.

In this connection it might be said that, although this system of construction has been employed before, the building under consideration is the largest example of this kind of work yet done in the neighborhood of New York City. It was adopted instead of corrugated iron, as it is much more substantial, and it was considered preferable to brick, as the later would have required much more extensive foundations.

The doors at each of the bays of the building are of rolling steel shutter type, and are composed of rolled-steel strips which interloop with each other, so that while the entire door is of steel, it can easily be raised and lowered.

Capacity and Pit Room

All of the tracks in the north and middle bays are supplied with pits for inspecting purposes, and as each track has a length sufficient to hold six cars, the capacity of these two bays is fifty-four cars.

The inspection pits are heated by steam and lighted by electric light, for which latter purpose frequent sockets are provided, and are also equipped with gas pipes, so that gas torches can be used instead of gasoline.

Trolley Connection

As usual in shops of this kind, the third rail is not carried into the shops, but the cars will be moved about by means of a special trolley. In the middle bay this trolley consists of a four-wheeled light-frame carriage, which will run on a conductor located in the pit. The carriage has attached to it a flexible wire which can be connected to the shoe-hanger of the truck or to the end plug of the car, so that the cars can be moved around in the shops by means of their own motors. In the north bay, where the pits are very shallow, the conductor is carried overhead and consists of an 8-pound T-rail supported from the roof girders.

The middle bay is provided with a 50-ton electric crane, which spans all of the tracks in this shop and is so arranged that it can serve any one of the thirty cars on the five tracks, and can deliver the trucks, wheels, motors, and other repair parts at either end of the shops, where they can be transferred to the telpherage hoist.

The Telpherage System

One of the most interesting features of the shops is the electric telpherage system. This system runs the entire length of the north and south bays crossing the middle bay or erection shop at each end, so that the telpherage hoist can pick up in the main room any wheels, trucks, or other apparatus which may be required, and can take them either into the north bay for painting, or into the south bay or machine shop for machine-tool work. The telpherage system extends across the transfer table pit at the west end of the shops and into the storehouse and blacksmith shop at the Seventh Avenue end of the grounds.

The traveling telpherage hoist has a capacity of 6,000 pounds. The girders upon which it runs consist of 12-inch I-beams, which are hung from the roof trusses. The car has a weight of one ton and is supported by and runs on the I-beam girders by means of four 9-inch diameter wheels, one on each side. The hoist is equipped with two motors. The driving motor of two horse power is geared by double reduction gearing to the driving wheels at one end of the hoist. The hoist motor is of eight horse power, and is connected by worm gearing and then by triple reduction gearing to the hoist drum. The motors are controlled by rheostatic controllers, one for each motor. The hoist motor is also fitted with an electric brake by which, when the power is cut off, a band brake is applied to the hoisting drum. There is also an automatic cut-out, consisting of a lever operated by a nut, which travels on the threaded extension of the hoisting drum shaft, and by which the current on the motor is cut off and the brake applied if the chain hook is wound up too close to the hoist.

Heating and Lighting

The buildings are heated throughout with steam, with vacuum system of return. The steam is supplied by two 100 horse power return tubular boilers, located at the southeastern corner of the building and provided with a 28-inch stack 60 feet high. The heat is distributed at 15 pounds pressure throughout the three bays by means of coil radiators, which are placed vertically against the side walls of the shop and storeroom. In addition, heating pipes are carried through the pits as already described. The shops are well lighted by large windows and skylights, and at night by enclosed arc lights.

INTERIOR VIEW OF 148TH STREET REPAIR SHOPS

Fire Protection

The shops and yards are equipped throughout with fire hydrants and fire plugs, hose and fire extinguishers. The water supply taps the city main at the corner of Fifth Avenue and 148th Street, and pipes are carried along the side of the north and south shops, with three reel connections on each line. A fire line is also carried through the yards, where there are four hydrants, also into the general storeroom.

General Store Room

The general storeroom, oil room, and blacksmith shop occupy a building 199 feet by 22 feet in the southwestern corner of the property. This building is of the same general construction as that of the inspection shops. The general storeroom, which is that fronting on 148th Street, is below the street grade, so that supplies can be loaded directly onto the telpherage hoist at the time of their receipt, and can be carried to any part of the works, or transferred to the proper compartments in the storeroom. Adjoining the general room is the oil and paint storeroom, which is separated from the rest of the building by fire walls. This room is fitted with a set of eight tanks, each with a capacity of 200 gallons. As the barrels filled with oil and other combustible material are brought into this room by the telpherage system they are deposited on elevated platforms, from which their contents can be tapped directly into the tank.

Blacksmith Shop

The final division of the west shops is that in the northeastern corner, which is devoted to a blacksmith shop. This shop contains six down-draught forges and one drop-hammer, and is also served by the telpherage system.

Transfer Table

Connecting the main shops with the storeroom and blacksmith or west shops is a rotary transfer table 46 feet 16-13/16 inches long and with a run of 219 feet. The transfer table is driven by a large electric motor the current being supplied through a conductor rail and sliding contact shoe. The transfer table runs on two tracks and is mounted on 33-inch standard car wheels.

Employees

The south side of the shop is fitted with offices for the Master Mechanic and his department.

The working force will comprise about 250 in the shops, and their lockers, lavatories, etc., are located in the south bay.

CHAPTER XII

SUB-CONTRACTORS

The scope of this book does not permit an enumeration of all the sub-contractors who have done work on the Rapid Transit Railroad. The following list, however, includes the sub-contractors for all the more important parts of the construction and equipment of the road.

General Construction, Sub-section Contracts, Track and Track Material, Station Finish, and Miscellaneous Contracts

S. L. F. DEYO, Chief Engineer.

Sub-sections

For construction purposes the road was divided into sub-sections, and sub-contracts were let which included excavation, construction and re-construction of sub-surface structures, support of surface railway tracks and abutting buildings, erection of steel (underground and viaduct), masonry work and tunnel work under the rivers; also the plastering and painting of the inside of tunnel walls and restoration of street surface.

Bradley, William, Sub-sections 6A and 6B, 60th Street to 104th Street.

Degnon-McLean Contracting Company (Degnon Contracting Company), Sub-section 1, 2 and 5A, Post-office to Great Jones Street and 41st Street and Park Avenue to 47th Street and Broadway.

Farrell, E. J., Sub-section, Lenox Avenue Extension, 142d Street to 148th Street.

Farrell & Hopper (Farrell, Hopper & Company), Sub-sections 7 and 8, 103d Street and Broadway to 135th Street and Lenox Avenue.

Holbrook, Cabot & Daly (Holbrook, Cabot & Daly Contracting Company), Sub-section 3, Great Jones Street to 33d Street.

McCabe & Brother, L. B. (R. C. Hunt, Superintendent), Sub-sections 13 and 14, 133d Street to Hillside Avenue.

McMullen & McBean, Sub-section 9A, 135th Street and Lenox Avenue to Gerard Avenue and 149th Street.

Naughton & Company (Naughton Company), Sub-section 5B, 47th Street to 60th Street.

Roberts, E. P., Sub-sections 10, 12, and 15, Foundations (Viaducts), Brook Avenue to Bronx Park, 125th Street to 133d Street, and Hillside Avenue to Bailey Avenue.

Rodgers, John C., Sub-section 9B, Gerard Avenue to Brook Avenue.

Shaler, Ira A. (Estate of Ira A. Shaler), Sub-section 4, 33d Street to 41st Street.

Shields, John, Sub-section 11, 104th Street to 125th Street.

Terry & Tench Construction Company (Terry & Tench Company), Sub-sections 10, 12, and 15, Steel Erection (Viaducts), Brook Avenue to Bronx Park, 125th Street to 133d Street, and Hillside Avenue to Bailey Avenue.

BROOKLYN EXTENSION.

Cranford & McNamee, Sub-section 3, Clinton Street to Flatbush and Atlantic Avenues, Brooklyn.

Degnon-McLean Contracting Company (Degnon Contracting Company), Sub-section 1, Park Row to Bridge Street, Manhattan.

Onderdonk, Andrew (New York Tunnel Company), Sub-sections 2 and 2A, Bridge Street, Manhattan, to Clinton and Joralemon Streets, Brooklyn.

TRACK AND TRACK MATERIAL

American Iron & Steel Manufacturing Company, Track Bolts.

Baxter & Company, G. S., Ties.

Connecticut Trap Rock Quarries, Ballast.

Dilworth, Porter & Company, Spikes.

Holbrook, Cabot & Rollins (Holbrook, Cabot & Rollins Corporation), Track Laying, City Hall to Broadway and 42d Street.

Long Clove Trap Rock Company, Ballast.

Malleable Iron Fittings Company, Cup Washers.

Naughton Company, Track Laying, Underground Portion of Road north of 42d Street and Broadway.

Pennsylvania Steel Company, Running Rails, Angle Bars, Tie Plates and Guard Rails.

Ramapo Iron Works, Frogs and Switches, Filler Blocks and Washers.

Sizer & Company, Robert R., Ties.

Terry & Tench Construction Company (Terry & Tench Company), Timber Decks for Viaduct Portions, and Laying and Surfacing Track on Viaduct Portions.

Weber Railway Joint Manufacturing Company, Weber Rail Joints.

STATION FINISH

American Mason Safety Tread Company, Safety Treads.

Atlantic Terra Cotta Company, Terra Cotta.

Boote Company, Alfred, Glazed Tile and Art Ceramic Tile.

Byrne & Murphy, Plumbing, 86th Street Station.

Dowd & Maslen, Brick Work for City Hall and other Stations and Superstructures for 72d Street, 103d Street and Columbia University Stations.

Empire City Marble Company, Marble.

Grueby Faience Company, Faience.

Guastavino Company, Guastavino Arch, City Hall Station.

Hecla Iron Works, Kiosks and Eight Stations on Elevated Structure.

Herring-Hall-Marvin Safe Company, Safes.

Holbrook, Cabot & Rollins Corporation, Painting Stations.

Howden Tile Company, Glazed Tile and Art Ceramic Tile.

Laheny Company, J. E., Painting Kiosks.

Manhattan Glass Tile Company, Glass Tile, and Art Ceramic Tile.

Parry, John H., Glass Tile and Art Ceramic Tile.

Pulsifer & Larson Company, Illuminated Station Signs.

Rookwood Pottery Company, Faience

Russell & Irwin Manufacturing Company, Hardware

Simmons Company, John, Railings and Gates.

Tracy Plumbing Company, Plumbing.

Tucker & Vinton, Strap Anchors for Kiosks.

Turner Construction Company, Stairways, Platforms, and Platform Overhangs.

Vulcanite Paving Company, Granolithic Floors.

MISCELLANEOUS

American Bridge Company, Structural Steel.

American Vitrified Conduit Company, Ducts.

Blanchite Process Paint Company, Plaster Work and Blanchite Enamel Finish on Tunnel Side Walls.

Brown Hoisting Machinery Company, Signal Houses at Four Stations.

Camp Company, H. B., Ducts.

Cunningham & Kearns, Sewer Construction, Mulberry Street, East 10th Street, and East 22d Street Sewers.

Fox & Company, John, Cast Iron.

McRoy Clay Works, Ducts.

Norton & Dalton, Sewer Construction, 142d Street Sewer.

Onondaga Vitrified Brick Company, Ducts.

Pilkington, James, Sewer Construction, Canal Street and Bleecker Street Sewers.

Simmons Company, John, Iron Railings, Viaduct Sections.

Sicilian Asphalt Paving Company, Waterproofing.

Tucker & Vinton, Vault Lights.

United Building Material Company, Cement.

Electrical Department

L. B. STILLWELL, Electrical Director.

Electric plant for generation, transmission, conversion, and distribution of power, third rail construction, electrical car equipment, lighting system, fire and emergency alarm systems:

American Steel & Wire Company, Cable.

Bajohr, Carl, Lightning Rods.

Broderick & Company, Contact Shoes.

Cambria Steel Company, Contact Rail.

Columbia Machine Works & Malleable Iron Company, Contact Shoes.

Consolidated Car Heating Company, Car Heaters.

D. & W. Fuse Company, Fuse Boxes and Fuses.

Electric Storage Battery Company, Storage Battery Plant.

Gamewell Fire Alarm Telegraph Company, Fire and Emergency Alarm Systems.

General Electric Company, Motors, Power House and Sub-station Switchboards, Control Apparatus, Cable.

General Incandescent Arc Light Company, Passenger Station Switchboards.

India Rubber & Gutta Percha Insulating Company, Cables.

Keasby & Mattison Company, Asbestos.

Malleable Iron Fittings Company, Third Rail and other Castings.

Mayer & Englund Company, Rail Bonds.

Mitchell Vance Company, Passenger Station Electric Light Fixtures.

National Conduit & Cable Company, Cables.

National Electric Company, Air Compressors.

Nernst Lamp Company, Power Station Lighting.

Okonite Company, Cables.

Prometheus Electric Company, Passenger Station Heaters.

Roebling's Sons Company, J. A., Cables.

Reconstructed Granite Company, Third Rail Insulators.

Standard Underground Cable Company, Cables.

Tucker Electrical Construction Company, Wiring for Tunnel and Passenger Station Lights.

Westinghouse Electric & Manufacturing Company, Alternators, Exciters, Transformers, Motors, Converters, Blower Outfits.

Westinghouse Machine Company, Turbo Alternators.

Mechanical and Architectural Department

John Van Vleck, Mechanical and Construction Engineer.

Power house and sub-station, steam plant, repair shop, tunnel drainage, elevators.

POWER HOUSE

Alberger Condenser Company, Condensing Equipment.

Allis-Chalmers Company, Nine 8,000-11,000 H. P. Engines.

Alphons Custodis Chimney Construction Company, Chimneys.

American Bridge Company, Structural Steel.

Babcock & Wilcox Company, Fifty-two 600 H. P. Boilers and Six Superheaters.

Burhorn, Edwin, Castings.

Gibson Iron Works, Thirty-six Hand-fired Grates.

Manning, Maxwell & Moore, Electric Traveling Cranes and Machine Tools.

Milliken Brothers, Ornamental Chimney Caps.

Otis Elevator Company, Freight Elevator.

Peirce, John, Power House Superstructure.

Power Specialty Company, Four Superheaters.

Ryan & Parker, Foundation Work and Condensing Water Tunnels, etc.

Robins Conveying Belt Company, Coal and Ash Handling Apparatus.

Reese, Jr., Company, Thomas, Coal Downtake Apparatus, Oil Tanks, etc.

Riter-Conley Manufacturing Company, Smoke Flue System.

Sturtevant Company, B. F., Blower Sets.

Tucker & Vinton, Concrete Hot Wells.

Treadwell & Company, M. H., Furnace Castings, etc.

Walworth Manufacturing Company, Steam, Water, and Drip Piping.

Westinghouse, Church, Kerr & Company, Three Turbo Generator Sets and Two Exciter Engines.

Westinghouse Machine Company, Stokers.

Wheeler Condenser Company, Feed Water Heaters.

Worthington, Henry R., Boiler Feed Pumps.

SUB-STATIONS

American Bridge Company, Structural Steel.

Carlin & Company, P. J., Foundation and Superstructure, Sub-station No. 15 (143d Street).

Cleveland Crane & Car Company, Hand Power Traveling Cranes.

Crow, W. L., Foundation and Superstructure Sub-stations Nos. 17 and 18 (Fox Street, Hillside Avenue).

Parker Company, John H., Foundation and Superstructure Sub-stations Nos. 11, 12, 13, 14, and 16 (City Hall Place, E. 19th Street, W. 53d Street, W. 96th Street, W. 132d Street).

INSPECTION SHED

American Bridge Company, Structural Steel.

Beggs & Company, James, Heating Boilers.

Elektron Manufacturing Company, Freight Elevator.

Farrell, E. J., Drainage System.

Hiscox & Company, W. T., Steam Heating System.

Leary & Curtis, Transformer House.

Milliken Brothers, Structural Steel and Iron for Storehouse.

Northern Engineering Works, Electric Telpherage System.

O'Rourke, John F., Foundation Work.

Tucker & Vinton, Superstructure of Reinforced Concrete.

Tracy Plumbing Company, Plumbing.

Weber, Hugh L., Superstructure of Storehouse, etc.

SIGNAL TOWERS

Tucker & Vinton, Reinforced Concrete Walls for Eight Signal Towers.

PASSENGER ELEVATORS

Otis Elevator Company, Electric Passenger Elevators for 167th Street, 181st Street, and Mott Avenue Stations, and Escalator for Manhattan Street Station.

Rolling Stock and Signal Department

GEORGE GIBBS, Consulting Engineer.

Cars, Automatic Signal System.

American Car & Foundry Company, Steel Car Bodies and Trailer Trucks.

Buffalo Forge Company, Blacksmith Shop Equipment.

Burnham, Williams & Company (Baldwin Locomotive Works), Motor Trucks.

Cambria Steel Company, Trailer Truck Axles.

Christensen Engineering Company, Compressors, Governors, and Pump Cages on Cars.

Curtain Supply Company, Car Window and Door Curtains.

Dressel Railway Lamp Works, Signal Lamps.

Hale & Kilburn Manufacturing Company, Car Seats and Backs.

Jewett Car Company, Wooden Car Bodies.

Manning, Maxwell & Moore, Machinery and Machine Tools for Inspection Shed.

Metal Plated Car & Lumber Company, Copper Sheathing for Cars.

Pitt Car Gate Company, Vestibule Door Operating Device for Cars.

Pneumatic Signal Company, Three Mechanical Interlocking Plants.

Standard Steel Works, Axles and Driving Wheels for Motor and Trailer Trucks.

St. Louis Car Company, Wooden Car Bodies and Trailer Trucks.

Stephenson Company, John, Wooden Car Bodies.

Taylor Iron & Steel Company, Trailer Truck Wheels.

Union Switch & Signal Company, Block Signal System and Interlocking Switch and Signal Plants.

Van Dorn Company, W. T., Car Couplings.

Wason Manufacturing Company, Wooden Car Bodies and Trailer Trucks.

Westinghouse Air Brake Company, Air Brakes.

Westinghouse Traction Brake Company, Air Brakes.

9 789356 784802